**AMERICAN COLLEGE
of SPORTS MEDICINE.**
w w w . a c s m . o r g

ACSM's
METABOLIC
CALCULATIONS
HANDBOOK

D0878434

EDITORS

Stephen Glass, PhD, FACSM

Professor

Grand Valley State University

Allendale, Michigan

Gregory B. Dwyer, PhD, FACSM

Associate Professor

East Stroudsburg University of Pennsylvania

East Stroudsburg, Pennsylvania

ACSM's
METABOLIC
CALCULATIONS
HANDBOOK

AMERICAN COLLEGE
of SPORTS MEDICINE
www.acsm.org

Lippincott Williams & Wilkins
a Wolters Kluwer business
Philadelphia · Baltimore · New York · London
Buenos Aires · Hong Kong · Sydney · Tokyo

Acquisitions Editor: Emily J. Lupash
Managing Editor: Matthew J. Hauber
Marketing Manager: Christen D. Murphy
Production Editor: Jennifer W. Glazer
Designer: Risa J. Clow
Compositor: International Typesetting and Composition
Printer: R.R. Donnelley & Sons—Crawfordsville

ACSM Publications Committee Chair: Jeffrey L. Roitman, EdD, FACSM
ACSM Group Publisher: D. Mark Robertson

Lippincott Williams & Wilkins
351 West Camden Street
Baltimore, MD 21201

530 Walnut Street
Philadelphia, PA 19106

To purchase additional copies of this book, call our customer service department at **(800) 638-3030** or fax orders to **(301) 223-2320.** International customers should call **(301) 223-2300.**

Visit Lippincott Williams & Wilkins on the Internet: http://www.LWW.com. Lippincott Williams & Wilkins customer service representatives are available from 8:30 am to 6:00 pm, EST.

For more information concerning American College of Sports Medicine certification and suggested preparatory materials, call **(800) 486-5643** or visit the American College of Sports Medicine web site **www.acsm.org.**

06 07 08 09 10
1 2 3 4 5 6 7 8 9 10

There is both an art and a science to exercise prescription. Individual differences require that exercise professionals be able to create specific, individualized programs for their clients. While mode, duration, and frequency are often easily chosen and adjusted, exercise intensity is something that requires specific tools to accurately quantify. The metabolic equations are valuable tools to assist the exercise professional in quantifying exercise intensity and caloric expenditure. They are also very helpful when setting the exercise intensity goal for the client. The ability to provide exercise intensity information that relates to specific measurements from a graded exercise test or some other functional capacity guideline means that clients can exercise in a safer and more prudent intensity range.

This text is designed for the exercise professional who may work in clinical as well as health-fitness settings. It is also designed for the instructor of a course in exercise physiology or exercise testing and prescription who is teaching the use of these equations. In the clinical setting, training intensity is often a very narrow range, and exceeding prescribed intensity can lead to adverse signs and symptoms. Data from an exercise test, in concert with the metabolic equations, can be used to set the proper training intensity. In many rehabilitation settings, exercise mode is often rotated to allow for a wider range of training adaptations. The metabolic equations are very helpful in setting appropriate work rates across modes. Similarly, in the health and fitness settings, exercise prescriptions are often based upon exercise test data, and the metabolic equations will be essential to help define the appropriate exercise intensity. For home-based exercise programs, speed and time can be measured from walking or running, and oxygen cost and caloric expenditure can be measured. For individuals tracking energy expenditure (i.e., weight loss program), the equations will help them determine their energy expenditure.

This text will also be useful for individuals working within rehabilitation settings where their primary goal may be injury rehabilitation. Often during rehabilitation from an injury, the clinician is also involved with training the whole person. For example, weight loss may alleviate some orthopedic pain symptoms, so the rehabilitation specialist may employ an exercise program. In summary, anyone who needs to set appropriate exercise intensity, or quantify the intensity that someone is already using, then this text is for you.

The metabolic equations provide oxygen consumption and work rate information, depending on how you complete them. Therefore the initial chapters cover the basics of metabolism and energy, explaining how oxygen can be used to determine caloric cost of an activity. The equations themselves use basic algebraic principles, and therefore a primer on basic algebraic is also included. The subsequent chapters are devoted to a step-by-step approach to understanding the equations. The equations can be used two ways: you can use work information (i.e., speed, grade, work rate, step rate) and calculate $\dot{V}O_2$, or if you know the desired $\dot{V}O_2$ for training you can calculate the appropriate work. We refer to this as working the equations forward or backward, and again the calculations are basic algebraic transformations.

Chapters 5–9 address each of the metabolic equations separately. The chapters begin with a basic introduction to the derivation of the equation, and provide examples of the various ways the equation can be used. Each chapter has a practice table with work intensity information missing. We anticipate this being a useful tool for the student and instructor in that the problem sets have been designed so that the student must work the equation both forward and backward in order to fill in the table. Following the table the answers are given as well as the step by step solutions. While there may be different approaches to solving the equations, we feel we have chosen the most straightforward for the non-mathmatically inclined. Chapter 10 is a summary chapter in which a set of problems are provided using all of the equations, solving both forward and backward. Step-by-step solutions to each is given. Chapter 11 is a test with only the solutions provided. The appendix provides a "cheat sheet" of sorts, which the student might use in class to assist with solving the problems. All of the equations as well as common conversions are given.

Our hope is that by going through the chapters, working the problems sets and following along with the step-by-step solutions, the student or professional will develop a strong familiarity with the practical uses of the metabolic equations.

ACKNOWLEDGMENTS

We would like to acknowledge to assistance of the reviewers who provided commentary and help in developing the chapters. Specifically, Mr. Clinton Brawner and Dr. Jacalyn Macomb provided extensive editing and comment that assisted with the format and wording. In addition, I would like to thank students at Grand Valley State University who were kind enough to provide feedback on some of the chapters (OK, so I gave it to them as an assignment). The students found an amazing number of typos and small errors; we are grateful. We would also like to acknowledge Dr. Mark Freitag of East Stroudsburg University for his assistance with the Math Primer chapter, and Dr. Charlene Beckman of Grand Valley State University for her assistance with formatting the math solutions. Finally, we would like to acknowledge Dr. Lenny Kaminsky of Ball State University for involving us in the development of the CD-ROM tutorial that was developed in 2000. Without his guidance we wouldn't be creating this text.

This text was reviewed for the American College of Sports Medicine by an expert review panel of the ACSM Committee on Certification and Registry Boards including:

Clinton A. Brawner, ACSM Exercise Specialist®
Henry Ford Hospital
Preventive Cardiology
Detroit, Michigan

Jacalyn J. McComb, PhD, FACSM, ACSM Exercise Specialist®
Texas Tech University
Lubbock, Texas

CONTENTS

INTRODUCTION

The importance of physical activity and/or exercise has been clearly demonstrated in numerous scientific reports and announcements for improving overall health and physical fitness. While it is universally agreed that physical activity is important for health, the amount of physical activity necessary to attain these health benefits has been debated. A more exacting exercise prescription or program from a health/fitness professional may be possible and even more desirable for many clients when they attempt to increase their physical activity. Health/fitness professionals may need to be able to calculate, by measuring or estimating, the energy requirements of various forms of physical activity to best advise their clients and to individualize the amount and type of physical activity needed to improve and maintain their client's health.

Energy requirements can be expressed in terms of the oxygen requirements of the physical activity being performed—commonly referred to as the oxygen consumption or oxygen cost ($\dot{V}O_2$). $\dot{V}O_2$ is perhaps best known as a maximal measure ($\dot{V}O_{2max}$) and provides useful information for health/fitness professionals of their client's cardiorespiratory fitness (CRF). However, under steady-state exercise conditions, $\dot{V}O_2$ provides a measure of the energy cost of physical activity, frequently expressed in kilocalories (kcals) and in combination with carbon dioxide production ($\dot{V}CO_2$) provides information about the *relative mixture* of metabolic substrates or fuel sources (carbohydrate versus fat; more about this in Chapter 2, Metabolic Primer) utilized.

Measured oxygen consumption ($\dot{V}O_2$), using open-circuit spirometry, provides the health/fitness professional with the best measure of the energy cost of physical activity. However, $\dot{V}O_2$ measurement may not be practical for nonclinical purposes as it is arduous and costly (this is covered in more detail in Chapter 2).

Therefore, the estimation of the energy requirements of physical activity is desirable. A popular method for the estimation of the energy requirements of physical activity employs the American College of Sports Medicine's (ACSM) Metabolic Calculations (ACSM MetCalc). The ACSM first introduced the ACSM MetCalc in 1975 to provide health/fitness professionals with a practical method to estimate the oxygen cost of common physical activities (walking, running, leg cycling, arm cycling, and stepping). The use of the ACSM MetCalc involves using basic algebraic principles (i.e., solving for the unknown), which has been the source of confusion for many aspiring health/fitness professionals in the past; hence the need for this written text. It should be noted that this written text follows the *ACSM's Guidelines for Exercise Testing and Prescription,* seventh edition, as well as the CD-ROM, *ACSM's Metabolic Calculations Tutorial,* version 1.0A.

Calculating the appropriate exercise workload needed to elicit the desired oxygen consumption or energy cost will allow the health/fitness professional to develop a more effective and individualized physical activity and/or exercise program for their client.

To sum up the need for the ACSM MetCalc, Table 1.1 presents a list of the reasons for this algebraic solution to the energy requirements for physical activity:

TABLE 1.1 Applications of the American College of Sports Medicine's Metabolic Calculations

- Oxygen cost ($\dot{V}O_2$) for several forms of physical activity can be estimated.
- Oxygen cost ($\dot{V}O_2$) can be easily transferred into energy cost (kcals).
- Estimating the rate of oxygen cost during physical activity allows for an estimate of the energy expenditure and hence caloric consumption associated with the activity.
- Exercise prescription and programming can be individualized to meet a client's needs and goals. Physical activity programs can be individualized for a client based upon their goals and needs and precise estimates for workloads (e.g., speed, grade, etc.) can be provided to achieve a certain level of metabolic stress.

2

METABOLIC PRIMER

Metabolism is the sum of all energy processes in the body, both the production of energy, as well as the use of energy. Thus, metabolism is all about energy: use and production. The use or breakdown of energy is a catabolic process while the buildup or production of energy is an anabolic process. Exercise and/or physical activity is associated with energy use, while nutrition deals with energy availability and storage. Energy usage can be expressed in terms of oxygen usage since the long term usage of energy in humans involves oxygen. The use of oxygen in metabolism is known as *aerobic metabolism* while the use of energy without the aid of oxygen is, by definition, *anaerobic metabolism*. The terms aerobic means with oxygen while anaerobic means without oxygen.

Metabolism is a function of time and intensity. Metabolism is a continuum from anaerobic to aerobic while anaerobic metabolism is more closely associated with quick, short-term, more intense activities that last up to about 1–2 minutes and aerobic metabolism is more likely to be utilized during activities that last more than 4 minutes and are typically lower in intensity. Thus, aerobic metabolism is more long term and involves the use of oxygen. Figure 2.1 below expresses the continuum of metabolism. Oxygen usage can be measured by using sophisticated methodology or can be estimated (given certain pre-conditions as discussed in the next chapter) with the use of the American College of Sports Medicine's Metabolic Calculations (ACSM MetCalcs).

CATABOLISM BASICS

Adenosine triphosphate (ATP) is the energy currency in the human body. Most of energy usage (from muscle contraction through digestion) involves the use of ATP. ATP can be produced in three ways: (1) from the storage depots of creatine phosphate and ATP; (2) from anaerobic glycolysis; and (3) from aerobic glycolysis. Creatine phosphate (CP) is easily converted to ATP in the skeletal muscle and represents an immediate form of energy directly in the active muscles that may last up to 30 seconds. This conversion of CP to ATP is known as the immediate energy

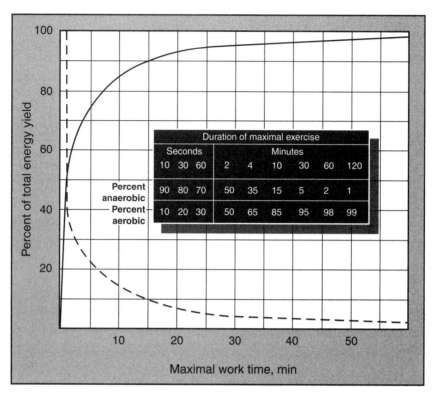

FIGURE 2.1 The Continuum of Metabolism. (Adapted with permission from McCardle WD, et al. *Essentials of Exercise Physiology* (2nd ed.). Philadelphia: Lippincott Williams & Wilkins, 2000.)

system and is anaerobic. However, the restoration of the CP and ATP stored in the muscle is aerobic.

Glycolysis is the conversion of glucose (or other simple carbohydrates) into ATP. Glycolysis may be either (1) anaerobic or (2) aerobic. Anaerobic glycolysis is a fast energy system that may last up to 2 minutes, does not involve oxygen in the energy pathway, and results in the production of the "end-product" lactic acid (that "tires" muscles). Aerobic metabolism is more complicated in its pathways (carbohydrate, fat, and amino acids may all participate as fuels) and may start from 2–4 minutes and continue indefinitely. Aerobic metabolism involves the use of oxygen.

During aerobic metabolism, either carbohydrate or fats are the major fuel substrates (amino acids from proteins are not utilized to a great extent). Aerobic metabolism is a slower energy system (therefore not able to produce ATP as quickly as is anaerobic glycolysis) but it is more complete in that the metabolic "leftovers," or end-products, of this pathway are carbon dioxide and water. Thus, aerobic metabolism consumes or uses oxygen and produces carbon dioxide.

EXPRESSIONS OF ENERGY USE

All actions in the human body require or use energy, from the digestion of food-stuff to muscle contraction. When we speak of exercise metabolism, we are often discussing the use of energy, known as energy expenditure. Energy expenditure in humans can be expressed several ways. Converting from one expression to another is relatively simple. To better understand energy expenditure you should be familiar with the following terms:

- **Aerobic metabolism:** production of energy using oxygen.
- **Oxygen consumption ($\dot{V}O_2$):** expression of the amount of oxygen used or consumed (typically as a rate or per minute).
- **Absolute versus relative:** relative, an expression relating the sum to some other value (such as body weight); while absolute, the expression of the value by itself.

Absolute Oxygen Consumption ($\dot{V}O_2$)

This is the volume of oxygen consumed by the person per unit (minute) of time, expressed in liters per minute ($L \cdot min^{-1}$) or milliliters per minute ($mL \cdot min^{-1}$). Resting absolute oxygen consumption ($\dot{V}O_{2rest}$) for a 70 kg person is approximately $0.25 \ L \cdot min^{-1}$. In highly trained individuals, maximal absolute oxygen consumption ($\dot{V}O_{2max}$) can be as high as $5.0 \ L \cdot min^{-1}$.

Absolute oxygen consumption is useful because it allows for an easy estimation of caloric expenditure. Each liter of O_2 consumed by the individual is associated with an energy expenditure of approximately 5 kilocalories (5 kcal). This will be discussed further in a later section.

Relative Oxygen Consumption ($\dot{V}O_2$)

This is the oxygen consumption relative to body weight, expressed in $mL \cdot kg^{-1} \cdot min^{-1}$; in other words, the volume of oxygen consumed by the cells of each kilogram of body weight every minute. The resting relative oxygen consumption ($\dot{V}O_{2rest}$) is approximately $3.5 \ mL \cdot kg^{-1} \cdot min^{-1}$. In highly trained aerobic athletes, a maximal relative oxygen consumption ($\dot{V}O_{2max}$) can be as high as $80.0 \ mL \cdot kg^{-1} \cdot min^{-1}$ (Figure 2.2).

In some instances, $\dot{V}O_2$ is expressed relative to a kilogram of fat-free mass, or to the square meters of body surface area, or to some other index of body size. Relative $\dot{V}O_2$ is commonly used to compare oxygen consumption of individuals who vary in body size. Because $\dot{V}O_{2max}$ is an index of cardiorespiratory fitness, a higher value is indicative of greater aerobic fitness. Cardiorespiratory fitness is defined in the most current edition of the *ACSM's Guidelines for Exercise Testing and Prescription* as the ability to perform large muscle, dynamic, moderate-to-high intensity exercise for prolonged periods.

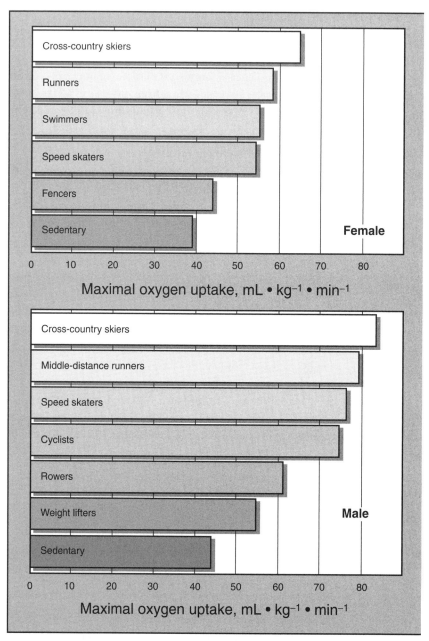

FIGURE 2.2 A Comparison of Relative Oxygen Consumptions. (Adapted with permission from McCardle WD, et al. *Essentials of Exercise Physiology (2nd ed.).* Philadelphia: Lippincott Williams & Wilkins, 2000.)

You can convert between the absolute and relative oxygen consumption using the following formulas:

$$\text{ABSOLUTE } (\text{mL} \cdot \text{min}^{-1}) \text{ to RELATIVE } (\text{mL} \cdot \text{kg}^{-1} \cdot \text{min}^{-1})$$

$$\frac{\dot{V}O_2(\text{mL} \cdot \text{min}^{-1})}{\text{Kilogram (kg) body mass}} = \text{mL} \cdot \text{kg}^{-1} \cdot \text{min}^{-1}$$

$$\text{RELATIVE to ABSOLUTE}$$

$$\frac{\dot{V}O_2(\text{mL} \cdot \text{kg}^{-1} \cdot \text{min}^{-1})}{1000} = \text{L} \cdot \text{kg}^{-1} \cdot \text{min}^{-1} \times \text{kg body mass} = \text{mL} \cdot \text{min}^{-1}$$

Metabolic Equivalents (METs)

Physicians and clinicians commonly use the term *MET(s)* as an expression of relative energy expenditure. A MET is an abbreviation for a metabolic equivalent. One MET is equivalent to the relative oxygen consumption at rest. Therefore, $1 \text{ MET} = 3.5 \text{ mL} \cdot \text{kg}^{-1} \cdot \text{min}^{-1}$. METs are calculated by dividing the relative oxygen consumption $(\text{mL} \cdot \text{kg}^{-1} \cdot \text{min}^{-1})$ by 3.5. For example, an individual consuming $35 \text{ mL} \cdot \text{kg}^{-1} \cdot \text{min}^{-1}$ of oxygen during steady-state physical activity is exercising at 10 METs. A MET is a useful expression because it allows for an easy comparison of the amount of oxygen consumption during exercise with that at rest. In addition, there is not any unit of measure associated with MET(s).

Calories

This expression of energy intake and expenditure is commonly used to quantify the amount of energy derived from foodstuffs, as well as the amount of energy expended at rest and during physical activity. A calorie is a very small unit. Most often the measurement kilocalories (kcal) is used. One kcal equals 1,000 calories. An average-sized individual who runs or walks one mile expends approximately 100 kcals. Also, a somewhat useful comparison for weight management is that every pound of fat contains about 3,500 kcals. The kcal can also be expressed in terms of a rate $(\text{kcal} \cdot \text{min}^{-1})$ and is then very useful in terms of the absolute energy intensity of exercise. For instance, one can express the exercise intensity in terms of $\text{kcal} \cdot \text{min}^{-1}$. An exercise of $10 \text{ kcal} \cdot \text{min}^{-1}$ means the individual is expending 10 kcal of energy for every minute they exercise:

$$10 \text{ kcals} \cdot \text{min}^{-1} = \frac{10 \text{ kcals} \cdot \text{min}^{-1}}{(5 \text{ kcal} \cdot \text{L}^{-1})} = 2 \text{ L} \cdot \text{min}^{-1} \text{ oxygen consumption}$$

Energy Stores

The respiratory exchange ratio may be used to help assess the relative use of carbohydrates versus fats in aerobic metabolism (assuming that the amino acids from proteins are not a major supplier of ATP for exercise catabolism). The respiratory exchange ratio (RER) is the ratio of the carbon dioxide produced to the oxygen consumed (RER = $\dot{V}CO_2/\dot{V}O_2$). When the RER is close to 0.70, fats are the primary fuel source for energy metabolism for the exercise. When the RER is closer to 1.0, then the primary fuel source for energy metabolism is carbohydrates.

CARBOHYDRATE STORES

Carbohydrates are stored in three main areas in the body: (1) in the muscle as glycogen, (2) in the liver as glycogen, and (3) in the blood as glucose. The total storage of carbohydrates in the body may be less than approximately 2,000 kcals.

FAT STORES

The human body stores the majority of excess energy intake as fat either below the skin as subcutaneous fat or around the internal organs. The amount of fat that is stored for energy usage can be represented by the body composition, or percent body fat. Even a lean individual may have in excess of 70,000 kcals of stored fat in their body. It takes approximately 3,500 calories to make and store 1 pound of body fat. Stated in reverse, 1 pound of fat can provide the body with 3,500 calories—the amount of energy needed to walk or run about 35 miles!

Net Versus Gross $\dot{V}O_2$

Gross $\dot{V}O_2$ refers to the total oxygen consumption, while net $\dot{V}O_2$ refers to the oxygen consumption for only the activity portion (or minus the resting component). All ACSM Metabolic Calculations (ACSM MetCalcs) provide gross $\dot{V}O_2$ values. In terms of exercise prescription, it may be helpful to use the net $\dot{V}O_2$ value; thus you may need to delete the resting component ($3.5 \text{ mL} \cdot \text{kg}^{-1} \cdot \text{min}^{-1}$). The concept of $\dot{V}O_2$ reserve used in exercise prescription involves the use of the net $\dot{V}O_2$ value ($\dot{V}O_2$ reserve is similar in concept to the heart rate reserve).

Net $\dot{V}O_2$ may also be used to assess the caloric cost of exercise. Net and gross oxygen consumption can be expressed in relative or absolute terms. The net $\dot{V}O_2$ is calculated by subtracting the resting $\dot{V}O_2$ from the gross $\dot{V}O_2$.

$$\text{Net } \dot{V}O_2 = \text{Gross } \dot{V}O_2 - \text{resting } \dot{V}O_2$$

- The gross rate of oxygen consumption is the total $\dot{V}O_2$ including the resting oxygen requirements, expressed as either $\text{L} \cdot \text{min}^{-1}$ or $\text{mL} \cdot \text{kg}^{-1} \cdot \text{min}^{-1}$.
- The net rate of oxygen consumption is the $\dot{V}O_2$ associated with only the amount of exercise being performed exclusive of resting oxygen uptake, expressed as either $\text{L} \cdot \text{min}^{-1}$ or $\text{mL} \cdot \text{kg}^{-1} \cdot \text{min}^{-1}$.

Measurement of Oxygen Consumption

The measurement of $\dot{V}O_2$ is typically performed in exercise laboratories or clinical settings using a procedure called open-circuit spirometry. During open-circuit spirometry the subject uses a mouthpiece and noseclip (or mask), which directs the expired air to an integrated metabolic system and computer interface that measures the total volume of air expired and percentage (%) of O_2 and CO_2 of the expired air. Subsequently, oxygen consumption ($\dot{V}O_2$) and carbon dioxide production ($\dot{V}CO_2$) are calculated. In this notation for $\dot{V}O_2$:

- The V stands for volume.
- The dot above the V ($\dot{V}$) denotes a rate, that is, the volume of oxygen consumed per unit of time, typically per minute.
- The O_2 stands for oxygen.

The measurement of $\dot{V}O_2$ is not always available or convenient, as has already been discussed, thus the need for the prediction of the $\dot{V}O_2$ with the ACSM MetCalcs.

3

MATHEMATICAL PRIMER

The ACSM MetCalcs equations require only a basic level of mathematical skills to solve. In this chapter, we will review these basic mathematical skills and hopefully alleviate any math-phobia you may have. Essentially, the mathematical discipline of algebra is used to solve for the ACSM MetCalcs. Algebra is one of the disciplines in math as is trigonometry and geometry (I bet you fondly remember these from your high school days). Algebra may be summarized as solving for the unknown in an equation. In the case of the ACSM MetCalcs, algebra is used to solve for an unknown metabolic cost (or $\dot{V}O_2$) based upon having some data on exercise workloads.

By use of a mathematical equation, we can solve for one value (y) based upon data for another value (x); this can be written mathematically as $y = mx + b$. (There are many ways to write equations; $y = mx + b$ represents the linear equation known as the y intercept/slope equation.) This equation, sure to bring back memories of math, can also be written as $y = a + bx$. While algebra may also be used to derive, or create, the equation, the ACSM MetCalcs give us the equation(s) to solve as opposed to having to derive them. In essence, we only need to know how to solve, or "work," the equations.

BASIC MATHEMATICAL OPERATIONS (SOLVING FOR THE UNKNOWN)

Here are some general mathematical examples for your consideration in solving a basic equation for the unknown.

Example 1. Solve for x in the following equation:

$$x - 3 = 7$$

Solution:

1. Add 3 to both sides of the equation (to get x by itself on one side of the equation):

$$x = 7 + 3$$

2. $x = 10$

Example 2. Solve for x in the following equation:

$$2x + 7 = 3$$

Solution:

1. Subtract 7 from both sides of the equation (to isolate x on one side of the equation):

$$2x + 7 - 7 = 3 - 7$$

2. Next, divide both sides by 2:

$$\frac{2x}{2} = \frac{-4}{2}$$

3. x = -2

Example 3: Solve for x in the following equation:

$$4x - \frac{3}{4} = \frac{7}{9}$$

Solution:

1. Add 3/4 to both sides of the equation (to isolate x on one side of the equation):

$$4x - \frac{3}{4} + \frac{3}{4} = \frac{7}{9} + \frac{3}{4}$$

$$4x = \frac{55}{36}$$

2. Next, divide both sides by 4:

$$\frac{4x}{4} = \frac{55/36}{4/1}$$

$$\frac{55/36}{4/1}$$ is the same as:

$$\frac{55}{36} \times \frac{1}{4} = \frac{55}{144}$$

3. $x = \dfrac{55}{144}$

Standard Algebraic Equation Format

$$y = mx + b$$

Algebraic equations predict a y value on basis of x data, where

y = energy cost ($\dot{V}O_2$)

m = slope of the prediction line. M is equal to the unit change in y, for every one unit change in x

x = workload (for instance, speed in meters per minute for walking or running)

b = a constant; resting energy expenditure (rest $\dot{V}O_2$). b also represents the intercept of the prediction line at the y axis (y intercept). There is a graph of this relationship in Figure 3.1.

Algebraic Equation Example

In the equation $y = 1.5x + 3$, we can solve for the value of y in this equation, when x is equal to 5 by the following:

1. $y = 1.5(5) + 3$ [first we multiply]
2. $y = 7.5 + 3$ [then we add]
3. $y = 10.5$ [to solve for the unknown y]

Of course, to solve for this you must remember your Mathematical Order of Operations from math class. As a quick review, the order in which you should perform certain mathematical operations is: (y^x, ×, /, +, −) from left to right in the equation. A way of remembering this is the acronym PEMDAS.

1. First, perform all operations inside any of the parenthesis; ( ); P
2. Next, perform all exponents; (y^x); E
3. Next, perform all multiplications; (x; ×); M
4. Next, perform all divisions: (÷; /); D
5. Next, perform all additions; (+); A
6. Finally, perform all subtractions; (−); S

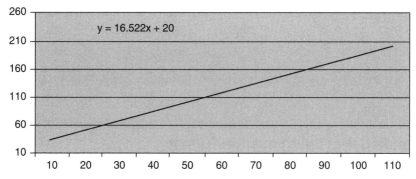

FIGURE 3.1 **The Simple Equation; y = mx + b Can Be Represented in the Following Graph.**
The solid line is represented by the equation $y = 16.522x + 20$, where $m = 16.522$ and $b = 20$.

More Complicated Algebraic Equations

The basic algebraic equation of $y = mx + b$ has two variables (y and x). However, an algebraic equation can have more than two variables ($x_1, x_2, \ldots$)

For example, an algebraic equation can be expressed as:

$$y = m_1x_1 + m_2x_2 + b$$

An example of solving for this algebraic equation follows:

Solve for y when $x_1 = 35$ and $x_2 = 10$ and $m_1 = 2$ and $m_2 = 1.5$ and $b = 3.5$, using the following equation ($y = m_1x_1 + m_2x_2 + b$):

1. First, all the values of x_1 and x_2 and m_1 and m_2 and b into the equation:

$$y = 2(35) + 1.5(10) + 3.5 \quad \text{or} \quad y = 2x_1 + 1.5x_2 + 3.5$$

2. Next, multiply x_1 and x_2 by m_1 and m_2 as below:

$$y = 70 + 15 + 3.5$$

3. Finally, add all three terms together for the answer:

$$y = 88.5$$

APPLICATION TO ACSM METABOLIC EQUATIONS

The formula format, $y = mx + b$, has one variable (x) to enter into the algebraic equation. This is the common format of the leg-cycling and arm-cycling metabolic equations. The leg-cycling equation is covered in Chapter 7; the arm-cycling equation is covered in Chapter 8.

The formula format, $y = m_1x_1 + m_2x_2 + b$, has two variables (x_1 and x_2) to enter into the algebraic equation. This is the common format of the walking and running metabolic equations. The walking equation is covered in Chapter 5; the running equation is covered in Chapter 6.

The stepping metabolic equation, while more complicated than the others, has only two variables to enter into the equation (step height and step frequency). The stepping equation is covered in more detail in Chapter 9.

ACSM METCALCS EQUATION SOLVING STEPS

The ACSM MetCalcs equations solving steps are as follow:

1. First, read the **ENTIRE** problem and decide exactly what is being asked. In other words, what is the question asking you to solve? Do not do any calculations until you have read and understand the entire question.

2. Identify the units that you wish the answer to be in: ($\dot{V}O_2$ for metabolic equations). More is written about this concept in the next chapter.

- $mL \cdot kg^{-1} \cdot min^{-1}$ (relative to an individual's body weight)
- METs (relative expression of oxygen uptake)
- $mL \cdot min^{-1}$ (absolute expression; independent of body weight)
- $L \cdot min^{-1}$ (absolute; no body weight)

As a note: x^{-1} is the scientific notation for "per" or for placing the unit under the division line. For instance, $mL \cdot kg^{-1} \cdot min^{-1}$ is the same as milliliters per kilogram per minute or milliliters/kilogram/minute (mL/kg/min). Also, the mathematical expression $x \cdot y$ is another way of writing $(x)(y)$ or x times y.

3. Identify the appropriate ACSM MetCalcs equation for the problem.
4. Convert all workloads, etc. to S.I. (Systeme International) units: S.I. units are similar to metric (this is discussed further below in Conversions section).
5. Isolate (by itself) the desired variable to solve (usually y) on one side of the equation.
6. Solve the equation (remember to apply the Mathematical Order of Operations: PEMDAS).

CONVERSIONS

Another important step to solving the ACSM MetCalcs is to convert all values to the appropriate units. While it may be more useful or common for us to discuss measurements in the units employed in United States, the ACSM MetCalcs use S.I. units in order that a unit of weight is expressed as pounds (lb) in the U.S. but across the world the preferred unit is kilogram (kg).

Some common conversion factors you need to remember/know for the ACSM MetCalcs are:

- Weight: pounds (lb) to kilograms (kg) [1 pound = 0.45 kg; 1 kg = 2.2 lb]
  For example, 145 lbs = 65.99 kg, or 70 kg = 154.37 lb.
- Height: inches (in) to centimeters (cm) [1 inch = 2.54 cm; 1 cm = 0.39 in]
  For example, 68 in = 172.72 cm, or 67 cm = 26.38 in.
- Speed: miles per hour (mph) to meters $\cdot$ min^{-1} [1 mph = 26.8 m $\cdot$ min^{-1}; 1 m $\cdot$ min^{-1} = 0.037 mph]

  For example, 3 mph = 80.4 m $\cdot$ min^{-1}, or 220 m $\cdot$ min^{-1} = 8.14 mph.

- Grade: % grade to decimal [1% = 1/100 or 0.10].
  For example, 5% grade on treadmill = 5/100 or 0.05.
- Power: kgm per minute to watts [1 kgm/min = 0.16 watt, 1 watt = 6.12 kgm/min]
  For example, 300 kgm/min = 49.02 watts, or 180 watts = 1101.6 kgm/min.

Conversions Applications: Arm- and Leg-Cycling

Most likely you may need to calculate workload on the cycle from the components of resistance, revolutions per minute, and meters per revolution. As an example, let's calculate the workload for cycle exercise using the popular Monark cycle.

Workload on cycle = kg of resistance times rpm (rev/min) times 6 m/rev on Monark cycle

$$= 2.0 \text{ kg} \times 50 \text{ rpm} \times 6\text{m/rev} = 600 \text{ kgm/min}$$

Workloads can be either expressed in kgm $\cdot$ min^{-1} or watts. Note that other cycle ergometers that may be used may have a different m/rev constant than the Monark. For instance, the Bodyguard cycle ergometer has a constant of 3 m/rev.

Conversions Applications: Walking and Running

Most likely you may need to convert Speed from miles per hour (miles/hr) to m $\cdot$ min^{-1}; also you may need to convert Grade from a percent (%) to a decimal (%/100).

ROUNDING OFF

It is best if you do not round off any of your calculations until the end answer is achieved. In other words, do not round off any of the intermediate steps until you have calculated the end result. When you do round off, it is best to round off to the hundredths place, or two places after the decimal (0.00). When you round off, you would round down for all integers from 1–4 and you would round up from integers from 5–9. For example:

- 10.3474 would be 10.35
- 7.272 would be 7.27

EXERCISE SCIENCE PREDICTION

It is often desirable in exercise science to predict or estimate a value as opposed to directly measuring that value for a number of reasons, such as the time it takes to perform the measurement, expense associated with the measurement, the inconvenience caused to the client in performing the measurement, etc. Some examples of prediction in exercise science include skinfold measurement to predict percent body fat, heart rate measurement during submaximal exercise to predict maximal exercise capacity ($\dot{V}O_{2max}$), and the ACSM MetCalcs.

Prediction Error

However, when we predict or estimate a value, we should understand and be willing to accept that there is some error in this prediction. Prediction error is an important part of all algebraic equations and thus, the ACSM Metcalcs.

Prediction error may be reported in several ways. The most common expression of prediction error is the standard error of the estimate (S.E.E.). The *ACSM's Guidelines for Exercise Testing and Prescription* make only limited references to error (and the S.E.E.) in the metabolic equations presented and this information is not typically a part of any multiple-choice questions used in certification exams.

Standard Error of the Estimate

The S.E.E. gives a range in which approximately two-thirds (67%) of the actual measurements are likely to fall within. When you consider the S.E.E., remember that approximately one-third (33%) of all individual actual measurements will be outside (greater than or less than) the range given by the S.E.E.

S.E.E. Graph

Figure 3.2 is a graph that demonstrates the concept of 1 S.E.E. above or below the line. This is a made-up graph of increasing oxygen consumption ($\dot{V}O_2$) with time as during a maximal exercise test. The *x* or horizontal axis represents time, in minutes while the *y* or vertical avis is $\dot{V}O_2$. The S.E.E. on the estimate for the prediction of $\dot{V}O_2$ is also contained on the graph. Remember, 1 S.E.E. above or below (written as $\pm$) represents 67% of all individuals, as discussed above.

S.E.E. Example

If the S.E.E. for the predicted $\dot{V}O_2$ from a metabolic equation is ± 3.5 mL $\cdot$ kg^{-1} $\cdot$ min^{-1}, then approximately 67% of all individuals will be within 3.5 mL $\cdot$ kg^{-1} $\cdot$ min^{-1} (plus or minus; $\pm$) of the predicted or estimated value. Another 33% of all individuals will be outside this range of ± 3.5 mL $\cdot$ kg^{-1} $\cdot$ min^{-1} of the predicted or estimated value.

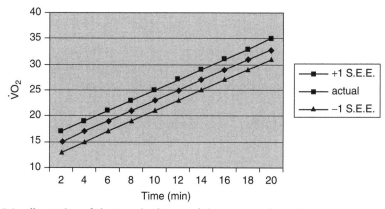

FIGURE 3.2 Illustration of the Standard Error of the Estimate (S.E.E.).

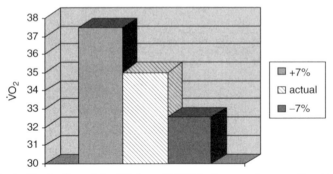

FIGURE 3.3 An Illustration of the 7% Error (S.E.E.) in Oxygen Consumption Prediction With the ACSM MetCalcs.

Another S.E.E. Example

It is also possible and desirable to express the S.E.E. as a percentage (i.e., ±7%). If the S.E.E. of a metabolic equation is ±7%, then the impact of this S.E.E. can be shown by multiplying the percentage by the predicted value (as seen in Figure 3.3). For example, if the predicted value for $\dot{V}O_2$ is 35 mL · kg^{-1} · min^{-1}, then the S.E.E. is ±2.45 mL · kg^{-1} · min^{-1}.

4

INTRODUCTION TO THE ACSM METABOLIC CALCULATIONS

PURPOSE

As has been discussed, the measurement of oxygen consumption requires equipment that is expensive and can be sophisticated; trained professional staff who can perform the assessment as well as interpret the data; and furthermore it does not lend itself to measuring large numbers of subjects. Therefore, in most non-laboratory or fitness situations, measuring $\dot{V}O_2$ is impractical. When measuring $\dot{V}O_2$ is not possible or feasible, reasonable estimates of $\dot{V}O_2$ during exercise can be made from regression equations derived from measured $\dot{V}O_2$ during steady-state exercise on ergometric devices and while walking or running. Thus, the purpose of the ACSM MetCalcs is to allow for estimation of energy cost (both oxygen consumption and caloric cost) from known workloads. Moreover, for exercise-programming purposes, these equations may be used to determine the required exercise workload or intensity associated with a desired level of energy expenditure.

HISTORY

The ACSM MetCalcs came to be known as such with the first edition of the *ACSM's Guidelines for Exercise Testing and Prescription (GETP)* published in 1975. The ACSM MetCalcs are equations that were gathered by several authors associated with the development of the *GETP* over the years from many scientific publications. In the sixth edition of *GETP* (2000), several equations were altered so that all five equations solve for relative oxygen consumption ($mL \cdot kg^{-1} \cdot min^{-1}$). There have been efforts by some to simplify these equations so as to use units for work typical in the United States (e.g., mph, % grade, etc.) as opposed to using the metric expressions for work (S.I.) employed (e.g., $m \cdot min^{-1}$, decimal grade, etc.). However, the ACSM MetCalcs have not 'adopted' this simplification.

BASIC SETUP

As was presented in the Mathematical Primer chapter (Chapter 3), the ACSM MetCalcs are generally set up as:

$$y = (m_1 x_1) + b \quad or \quad y = (m_1 x_1) + (m_2 x_2) + b$$

Where y = oxygen cost ($mL \cdot kg^{-1} \cdot min^{-1}$); m is a constant; x is an expression for work rate; and b represents resting oxygen consumption ($mL \cdot kg^{-1} \cdot min^{-1}$).

Note: for leg and arm cycling there is only one term (i.e., m_1x_1) while for tread-mill walking and running there are two terms (i.e., (m_1x_1) and (m_2x_2)).

The oxygen cost for physical activity may be expressed in absolute (i.e., $L \cdot min^{-1}$ or $mL \cdot min^{-1}$) or relative terms (i.e., $mL \cdot kg^{-1} \cdot min^{-1}$ or METs), as discussed in the Metabolic Primer chapter (Chapter 2). In addition, the oxygen cost of phys-ical activity may be expressed in net versus gross terms. Net oxygen consumption would denote the oxygen cost for physical activity only (without considering the resting oxygen consumption) while gross cost would include the cost of the activ-ity in addition to the resting oxygen cost. All ACSM MetCalcs yield gross oxygen consumption in relative terms ($mL \cdot kg^{-1} \cdot min^{-1}$).

Table D-1 in the seventh edition of *ACSM's Guidelines for Exercise Testing and Prescription* (reprinted later in this chapter) presents the metabolic equations for the gross or total oxygen consumption of walking, running, leg ergometry, arm ergometry, and stepping. For each prediction equation or ACSM MetCalc, there are some known physiological constants, such as how much oxygen is required to move the body horizontally (walking on the flat) and vertically (walking up a grade or hill) or the oxygen costs of pedaling at no resistance, that are used.

PUTTING IT ALL TOGETHER

In the Mathematical Primer chapter (Chapter 3), a step-by-step approach to solv-ing algebraic equations similar to the ACSM MetCalcs was presented. However, a few other, more specific points should be mentioned before you try to solve any problem using the ACSM MetCalcs. Using this systematic approach, the task of solving the problem is made much easier and will help you avoid mistakes.

1. Read each question carefully and do not proceed until you know what you are expected to calculate.
2. Extract the required information. Do not be misled with extraneous informa-tion. If, for example, a question wants you to calculate the $\dot{V}O_2$ for walking on a treadmill, volunteered data about the height, age, health history or the gen-der of the subject are irrelevant.
3. Select the correct metabolic equation. A common error committed by many candidates is choosing the wrong formula. If the subject was running on the treadmill at 5.5 miles/hr (5.5 mph × 26.8 $m \cdot min^{-1}$/miles/hr = 147.4 $m \cdot min^{-1}$) be sure to use the running equation (and not the walking equation). Be sure to check that the exercise conditions in the problem match the conditions of the ACSM MetCalcs (i.e., running is > 5.0 miles/hr; >134 $m \cdot min^{-1}$).
4. Write out each step. Do not take shortcuts.
5. Where needed, convert all values to the appropriate units (i.e., metric or S.I. units, such as kg, $m \cdot min^{-1}$, $kg \cdot m \cdot min^{-1}$).

6. Write down the formula and plug in the known values. Write clearly and place units after all variables. Be sure to consult Table D-1 from the most current ACSM GETP for the units.

7. Solve for the unknown. If the unknown is on the left side of the equation (i.e., the $\dot{V}O_2$ value), simply calculate the sum of all the components of the appropriate equation. If the unknown is on the right side of the equation, you will need to apply mathematical principles (as discussed in Chapter 3) to solve for the unknown.

8. Be sure to only round off the answer at the last step of the calculation and only round off to the hundredths (i.e., X.XX) place.

9. Examine the answer. Is the answer logical? Does it fall within the expected "normal" value or reasonable value?

10. Examine the choices. Make sure that your answer is in the same units as the answer on the examination, especially if a question does not specify what energy expression is needed (i.e., relative or absolute $\dot{V}O_2$, METs, kcal).

ASSUMPTIONS AND LIMITATIONS

The ACSM MetCalcs have been put together from several sources and have a common set of "conditions" that should be met in order for these calculations to be accurate. Assumptions are conditions that we must accept when using the equations (such as selected workloads), while limitations are issues that should be considered that affect the applicability of the equations (such as submaximal, steady-state aerobic exercise). Clearly a particular "condition" may be both an assumption and a limitation, thus we have chosen to discuss these as one single group. We have provided a brief discussion of several of these "conditions."

- The measured $\dot{V}O_2$ at a given work rate is highly reproducible for a given individual; that is, the $\dot{V}O_2$ at the same exercise intensity for the same individual will be very similar every time they exercise. However, the intersubject variability (variability between different subjects) in measured $\dot{V}O_2$ may have a standard error of estimate (S.E.E.) as high as 7%. A 7% S.E.E. may translate into a difference of up to or more than one MET ($3.5 \text{ mL} \cdot \text{kg}^{-1} \cdot \text{min}^{-1}$) in 67% (2/3) of individuals (see the discussion on S.E.E. in Chapter 3). Therefore, the equations work well if tracking the same subject over time, but are less accurate for comparing $\dot{V}O_2$ between different individuals and should be used with caution for this purpose.

- These equations were derived during steady-state submaximal aerobic exercise; therefore, they are only appropriate for predicting $\dot{V}O_2$ during steady-state submaximal aerobic exercise. The $\dot{V}O_2$ will be either overestimated or understimated when the contribution from anaerobic metabolism is large (such as near maximal exertion) or during non–steady-state exercise conditions. Remember, a ballpark figure for steady-state exercise, as taken from the submaximal cycle ergometer test, is a difference of less than or equal to 5 beats per minute ($\leq$5 beats/min) between successive minutes. When using the ACSM MetCalcs be sure that the exercise being conducted is both submaximal as well as steady-state.

- Although the accuracy of these equations is unaffected by most environmental influences (e.g., heat and cold), variables that change the mechanical efficiency (e.g., gait abnormalities, wind, snow, or sand) will result in a loss of accuracy.
- The use of the ACSM MetCalcs presupposes that ergometers used are calibrated properly and used appropriately (e.g., no rail-holding during treadmill exercise).
- The equations are most accurate at the stated speeds and power outputs. For instance, there is a range of walking/jogging/running speeds in which neither the walking nor the running equations are applicable. These speeds (3.7 to 5.0 miles/hr; 101–134 m $\cdot$ min^{-1}) are in the range of the transition from a walking to running motion or gait. The actual transition between walking and running in an individual varies depending on the individual's size, leg length, stride length, and normal walking pace. Therefore, inter-individual (between individuals) variability in $\dot{V}O_2$ is very wide within this speed range.

Despite these caveats, the proper and judicious use of the ACSM MetCalcs can provide valuable information to the health and fitness professional.

FORWARD AND BACKWARD SOLVING: FOR $\dot{V}O_2$ OR FOR WORKLOAD

The answer to any problem using the ACSM MetCalcs may be on either the right or left side of the equation. If we are trying to solve for oxygen cost or $\dot{V}O_2$ for the individual exercising at a particular workload on a particular piece of exercise equipment (i.e., walking at 2.5 miles/hr on a treadmill), the problem is a forward-solving one with the final answer on the left side of the equation, for example:

What is the oxygen cost ($\dot{V}O_2$) for a person walking at 2.5 miles/hr on a treadmill?

However, we may wish to calculate an appropriate workload on a piece of exercise equipment (i.e., treadmill speed). If so, then the problem is a backward-solving one with the final answer on the right side of the equation, for example:

At what workload (speed; assuming a zero grade) should a person walk on a treadmill to expend an oxygen cost ($\dot{V}O_2$) of 14.5 mL $\cdot$ kg^{-1} $\cdot$ min^{-1}?

Both these types of problems can be solved with the use of the ACSM MetCalcs. The forward-solving one is certainly more straightforward, with simple insertion of the relevant information (workload). The backward-solving one

TABLE D-1 **Metabolic Equations for Gross $\dot{V}O_2$ in Metric Units**

Walking

$\dot{V}O_2(mL \cdot kg^{-1} \cdot min^{-1}) = (0.1 \cdot S) + (1.8 \cdot S \cdot G) + 3.5\ mL \cdot kg^{-1} \cdot min^{-1}$

$\dot{V}O_2(mL \cdot kg^{-1} \cdot min^{-1}) = [0.1\ mL \cdot kg^{-1} \cdot meter^{-1} \cdot S(m \cdot min^{-1})]$
$+ [1.8\ mL \cdot kg^{-1} \cdot meter^{-1} \cdot S(m \cdot min^{-1}) \cdot G] + 3.5\ mL \cdot kg^{-1} \cdot min^{-1}$

Running

$\dot{V}O_2(mL \cdot kg^{-1} \cdot min^{-1}) = (0.2 \cdot S) + (0.9 \cdot S \cdot G) + 3.5\ mL \cdot kg^{-1} \cdot min^{-1}$

$\dot{V}O_2(mL \cdot kg^{-1} \cdot min^{-1}) = [0.2\ mL \cdot kg^{-1} \cdot meter^{-1} \cdot S(m \cdot min^{-1})]$
$+ [0.9\ mL \cdot kg^{-1} \cdot meter^{-1} \cdot S(m \cdot min^{-1}) \cdot G] + 3.5\ mL \cdot kg^{-1} \cdot min^{-1}$

Leg Cycling

$\dot{V}O_2(mL \cdot kg^{-1} \cdot min^{-1}) = 1.8\ (work\ rate)/(BM) + 3.5\ mL \cdot kg^{-1} \cdot min^{-1}$
$+ 3.5\ mL \cdot kg^{-1} \cdot min^{-1}$

$\dot{V}O_2(mL \cdot kg^{-1} \cdot min^{-1}) = (1.8\ mL \cdot kg^{-1} \cdot min^{-1})$
$\times (work\ rate\ in\ kg \cdot m \cdot min^{-1})(body\ mass\ in\ kg)$
$+ 3.5\ mL \cdot kg^{-1} \cdot min^{-1} + 3.5\ mL \cdot kg^{-1} \cdot min^{-1}$

Arm Cycling

$\dot{V}O_2(mL \cdot kg^{-1} \cdot min^{-1}) = 3\ (work\ rate)/(BM) + 3.5\ mL \cdot kg^{-1} \cdot min^{-1}$

$\dot{V}O_2(mL \cdot kg^{-1} \cdot min^{-1}) = (3\ mL \cdot kg^{-1} \cdot meter^{-1})$
$\times (work\ rate\ in\ kg \cdot m \cdot min^{-1})(body\ mass\ in\ kg) + 3.5\ mL \cdot kg^{-1} \cdot min^{-1}$

Stepping

$\dot{V}O_2(mL \cdot kg^{-1} \cdot min^{-1}) = (0.2 \cdot f) + (1.33 \cdot 1.8 \cdot H \cdot f) + 3.5\ mL \cdot kg^{-1} \cdot min^{-1}$

$\dot{V}O_2(mL \cdot kg^{-1} \cdot min^{-1}) = 0.2\ (steps \cdot min^{-1})$
$+ (1.33\ mL \cdot kg^{-1} \cdot meter^{-1})(1.8\ mL \cdot kg^{-1} \cdot meter^{-1})$
$\times (step\ height\ in\ meters)(steps \cdot min^{-1}) + 3.5\ mL \cdot kg^{-1} \cdot min^{-1}$

$\dot{V}O_2$ is gross oxygen consumption in $mL \cdot kg^{-1} \cdot min^{-1}$; S is speed in $m \cdot min^{-1}$; BM is body mass (kg); G is the percent grade expressed as a fraction; work rate $(kg \cdot m \cdot min^{-1})$; f is stepping frequency in minutes; H is step height in meters.

From *ACSM'S Guidelines for Excercise Testing and Prescription*, 7th ed. American College of Sports Medicine; Indiaopolis, 2006;289.

involves the use of higher mathematical skills with the simplification of the equation and the combination/elimination of like terms (see the Mathematical Primer chapter, Chapter 3, for review of this concept). You may or may not need to separate out the resting $\dot{V}O_2$, depending upon the problem. Thus, backward-solving involves more steps and careful reading/interpretation of the problem. Both examples of forward- and backward-solving of the ACSM MetCalcs will be presented in future chapters.

USE OF GETP APPENDIX TABLE D-1

Appendix Table D-1 from the seventh edition of the ACSM's GETP contains all the ACSM MetCalcs. As has been previously mentioned, all five of the ACSM MetCalcs solve for $\dot{V}O_2$ in relative terms ($mL \cdot kg^{-1} \cdot min^{-1}$). In addition, all the ACSM MetCalcs solve for $\dot{V}O_2$ in gross terms (the ACSM MetCalcs already include resting $\dot{V}O_2$). Each of the following chapters contain specific information on each of the ACSM MetCalcs (i.e., walking, running, leg ergometry, arm ergometry, and stepping). In the ACSM certification examination process, Table D-1 is provided for the candidates, thus eliminating the need to memorize the individual equations. However, knowing how to use each equation, is important. In addition, in the certification process, ACSM expects you to be able to perform some basic conversions. These conversions are listed below:

Walking/Running Speed:	Miles per hour (miles/hr) to meters $\cdot$ min^{-1} [1 mph = 26.8 m $\cdot$ min^{-1}]
Walking/Running Grade:	Percent (%) grade to decimal [1% =1/100 or 0.01]
Cycling Power Output:	kgm per minute to watts [1 kg $\cdot$ m $\cdot$ min^{-1} = 0.16 watt]
Body Mass:	Pounds (lb) to kilograms (kg) [1 pound = 0.45 kg]
Height:	Inches (in) to centimeters (cm) [1 inch = 2.54 cm]

Additionally, the following general comments are made about each individual ACSM MetCalc:

Walking: The constant for horizontal work for walking is 0.1 mL $\cdot$ kg^{-1} $\cdot$ min^{-1} per meters $\cdot$ min^{-1}. The constant for vertical work is 1.8 mL $\cdot$ kg^{-1} $\cdot$ min^{-1} per meters $\cdot$ min^{-1} per 1.0 grade. There is no need to consider an individual's body mass when calculating $\dot{V}O_2$.

Running: The constant for horizontal work for running is 0.2 mL $\cdot$ kg^{-1} $\cdot$ min^{-1}, per meters $\cdot$ min^{-1} (running involves lifting your body with an airborne phase). The constant for vertical work is 0.9 mL $\cdot$ kg^{-1} $\cdot$ min^{-1} per meters $\cdot$ min^{-1} per 1.0 grade. There is no need to consider an individual's body mass when calculating $\dot{V}O_2$.

Leg ergometer: Again, the constant for vertical work is 1.8 mL $\cdot$ kg^{-1} $\cdot$ min^{-1}, but this time per kg $\cdot$ m $\cdot$ min^{-1}. Body mass is required to calculate $\dot{V}O_2$. To account for resting $\dot{V}O_2$, both a resting $\dot{V}O_2$ (3.5 mL $\cdot$ kg^{-1} $\cdot$ min^{-1}) and a $\dot{V}O_2$ for unloaded cycling (3.5 mL $\cdot$ kg^{-1} $\cdot$ min^{-1}) is added. In addition, power output determination is specific to the type of ergometer. The Monark cycle's constant is 6.0 meters per revolution; this is multiplied by the revolutions traveled per minute (rev/min) and the resistance (kg) to determine workload or power output.

Arm ergometer: The constant for work is 3.0 mL $\cdot$ kg^{-1} $\cdot$ min^{-1} per kg $\cdot$ m $\cdot$ min^{-1}. Body mass is required to calculate $\dot{V}O_2$. In addition, power output determination is specific to the type of ergometer. The Monark Arm Ergometer™ is 2.5 meters per revolution; this is multiplied by the revolutions traveled per minute (rev/min) and the resistance (kg) to determine workload or power output.

Stepping: To determine the $\dot{V}O_2$, you need to know the frequency of stepping and the step height. Again, the vertical constant for work of 1.8 mL $\cdot$ kg^{-1} $\cdot$ min^{-1} per step height (m) and step frequency (meters $\cdot$ min^{-1}) is used. Body mass is not used to calculate $\dot{V}O_2$.

In the following chapters, we will examine more closely each ACSM MetCalc.

WALKING EQUATION

INTRODUCTION

Walking is a mode of exercise, that is unique in that it can be done anywhere, without equipment, and is helpful for enhancing cardiovascular fitness and weight management, as well as maintaining bone mineral density. The ACSM walking equation is a useful tool for determining proper training intensity and also can be used to estimate caloric expenditure during walking activities.

DERIVATION OF THE WALKING EQUATION

There are three components within the walking equation, each representing an aspect of energy expenditure. First, the oxygen cost of moving one kilogram of body weight one meter has been estimated to be 0.1 mL/kg/m. Therefore the **Horizontal Component** of walking can be computed as:

Horizontal Component = Speed $(m \cdot min^{-1}) \times 0.1$ mL/kg/m

Rewritten as:

$$= S \ (m \cdot min^{-1}) \times 0.1 \ mL/kg/m$$

The **Vertical Component** of walking can be calculated if we know the oxygen cost of moving vertically against gravity. This has been estimated to be 1.8 mL/min for each meter walked. Since the rate of movement (speed) as well as the steepness of the vertical climb (grade) also must be known, the Vertical Component contains all three of these components.

Vertical Component = Speed $(m \cdot min^{-1}) \times$ Grade (decimal) $\times 1.8$ mL/min/m

Rewritten as:

$$= S \ (m \cdot min^{-1}) \times G \times 1.8 \ mL/min/m$$

COMPUTING GRADE

Vertical ascent is denoted by grade, typically computed as a fraction (decimal), then converted to a percent. Percent grade reflects the degree of elevation gain for a given horizontal distance.

Example: A rise of 1 foot over a distance of 10 feet?

$$= 1 \text{ ft}/10 \text{ ft} = 0.10 \qquad 0.10 \cdot 100 = 10\% \text{ grade}$$

Note that when expressing grade within the walking equation, it is done in *decimal form:*

Example: 10% grade = **0.10**   5% grade = **0.05**

Together the horizontal and vertical components represent the net O_2 cost of walking, meaning the O_2 cost above rest. To obtain the gross $\dot{V}O_2$ we must add in the O_2 cost at rest (3.5 mL/kg/min). Resting metabolic rate has been computed at 1MET, or 3.5 mL/kg/min. This value is added in at the end of the equation.

THE ACSM WALKING EQUATION

The ACSM walking equation is:

$$\dot{V}O_2 \text{ (mL/kg/min)} = [\text{Speed (m/min)} \times 0.1 \text{ mL/kg/m}] + [\text{Speed (m/min)} \times \text{Grade (decimal)} \times 1.8 \text{ mL/min/m}] + 3.5 \text{ mL/kg/min}$$

Limitations of the Walking Equation

Limitations of the walking equation include the following:

1. *Steady-State Exercise Only!*
   The ACSM walking equation was computed based upon steady-state oxygen consumption values. Therefore, the equation will not be accurate when computing oxygen consumption during non–steady-state exercise. Using the ACSM walking equation to compute a maximal oxygen uptake (for example, the last stage of a maximal exercise test) will not be accurate during such non–steady-state exercise.
2. *Accuracy is dependent upon speed range!*
   The ACSM walking equation is only accurate between 1.9–3.7 miles per hour. Above the given speed range, walking economy changes, and individuals of differing height may actually run.

Outdoor verses Treadmill Walking

Research regarding the energy cost of outdoor walking compared to treadmill walking has shown that the ACSM walking equation is accurate and useful, regardless of the walking surface.

SPECIFIC USES OF THE ACSM WALKING EQUATION

The ACSM walking equation is typically utilized for two primary purposes. First, you can determine the oxygen cost of walking, either on level grade or an incline. The oxygen cost can then be used in determining the caloric expenditure for walking. Second, you can determine a specific walking speed and/or grade based upon a $\dot{V}O_2$ prescription.

Use 1: Determining the Oxygen Cost and Caloric Expenditure of Walking

DETERMINING OXYGEN COST

For individuals assigned a specific $\dot{V}O_2$ for training, the walking equation can help determine oxygen consumption, and thus quantify their work intensity.

Example: Bob was given an exercise prescription of 3–5 METs for his home exercise program. He is currently walking, and at the fitness center using a treadmill, walking 3.2 miles per hour up a 3% incline. Bob weighs 70 kg. Is Bob exercising in his prescribed zone?

Step 1: Use the ACSM walking equation to determine the oxygen cost of the activity.

$$\dot{V}O_2 \text{ (mL/kg/min)} = [S \text{ (m/min)} \cdot 0.1] + [S \text{ (m/min)} \cdot G \text{ (dec)} \cdot 1.8] + 3.5$$

Speed (S) conversion: 3.2 miles/hr $\cdot$ 26.8 = 85.76 m/min
Grade (G) conversion: 3%/100 = 0.03

$$\dot{V}O_2 \text{ (mL/kg/min)} = [85.76 \text{ m/min} \cdot 0.1] + [85.76 \text{ m/min} \cdot 0.03 \cdot 1.8] + 3.5 \text{ mL/kg/min}$$
$$= 8.58 \qquad\qquad + 4.63 \qquad\qquad\qquad + 3.5$$
$$= 16.71 \text{ mL/kg/min}$$

Step 2: Convert $\dot{V}O_2$ in mL/kg/min into METs (see Table 5.1 for conversion).

$$\frac{\dot{V}O_2 \text{ (mL/kg/min)}}{3.5}$$

$$\frac{16.71}{3.5} = 4.77 \text{ METs}$$

So, is Bob exercising within the prescribed range? Yes.

DETERMINING CALORIC COST

We can take the above results one step further, and estimate caloric expenditure as well. Caloric cost can be computed if we know the absolute $\dot{V}O_2$ in L/min (see Table 5.1).

Estimated Oxygen Cost: 16.71 mL/kg/min

Step 1: Convert $\dot{V}O_2$ in mL/kg/min into $\dot{V}O_2$ in mL/min. We simply need to multiply by body weight.

$$\dot{V}O_2 \text{ (mL/kg/min)} \cdot BW$$
$$16.71 \cdot 70$$
$$= 1169.7 \text{ mL/min}$$

Step 2: Convert $\dot{V}O_2$ in mL/min into $\dot{V}O_2$ in L/min. We simply need to divide by 1000!

$$\dot{V}O_2 \text{ (mL/min/1000)}$$
$$\frac{1169.7}{1000} = 1.17 \text{ L/min}$$

Step 3: Convert $\dot{V}O_2$ in L/min into kcals/min. We simply need to multiply $\dot{V}O_2$ L/min by 5.

$$\dot{V}O_2 \text{ L/min} \cdot 5$$
$$1.17 \cdot 5$$
$$= 5.85 \text{ kcals/min expended}$$

PRACTICE TABLE

PRACTICE TABLE 1. *Fill in the appropriate information and complete the table (answers start on page 32).*

Problem #	Walking speed (miles/hr)	Speed (m/min)	Walking grade (%)	Grade (decimal)	Body weight (kg)	$\dot{V}O_2$ mL/kg/min	$\dot{V}O_2$ mL/min	$\dot{V}O_2$ L/min	kcals/min
1	2.0		5		50				
2	3.4		0		90				
3	3.1		8		62				
4	1.7		10		75				

PRACTICE NOTES:

—*Did you convert speed from miles per hour to meters per minute?*
—*Did you convert grade from a percent to a fraction (decimal)?*
—*Remember, to convert from mL/kg/min to mL/min you must multiply by body weight.*

Use 2: Determine Speed and/or Grade Based Upon a $\dot{V}O_2$ Prescription

It is not uncommon for the initial phase of an exercise plan to involve some form of $\dot{V}O_{2max}$ assessment. Then, utilizing a specific percentage of $\dot{V}O_{2max}$ or $\dot{V}O_2$

reserve ($\dot{V}O_{2R}$), or simply a MET prescription, one can determine a specific walking intensity.

WALKING SPEED BASED UPON PERCENT $\dot{V}O_{2max}$

Example: Jane's $\dot{V}O_{2max}$ from a GXT was 21.5 mL/kg/min. As an initial intensity prescription, you choose to have her walk at 50% of her $\dot{V}O_{2max}$. Let's assume initially that she will walk without an incline, or 0% grade.

Step 1: Identify the training $\dot{V}O_2$.

50% of her $\dot{V}O_{2max}$ (21.5 mL/kg/min) is computed: $21.5 \cdot 0.50 = 10.75$ mL/kg/min.

Step 2: Using the training $\dot{V}O_2$, solve the walking equation to determine speed:

$$\dot{V}O_2 \text{ (mL/kg/min)} = [S \text{ (m/min)} \cdot 0.1] + [S \text{ (m/min)} \cdot G \text{ (dec)} \cdot 1.8] + 3.5$$

$$10.75 = [S \text{ m/min} \cdot 0.10] + [S \text{ m/min} \cdot 0 \cdot 1.8] + 3.5$$

$$-3.5 \qquad\qquad\qquad\qquad\qquad -3.5$$

$$7.25 = [S \text{ m/min} \cdot 0.10] + 0$$

$$\frac{7.25}{0.10} = [S \text{ m/min} \cdot \frac{0.10}{0.10}]$$

$$72.5 = S \text{ m/min}$$

Step 3: Convert speed in m/min to miles/hr.

Remember: 1 mile/hr = 26.8 m/min

$$\frac{72.5 \text{ m/min}}{26.8} = \textbf{2.7 miles/hr}$$

Answer: The walking speed to achieve 50% of $\dot{V}O_{2max}$ is 2.7 miles/hr

WALKING SPEED BASED UPON $\dot{V}O_2$ RESERVE

Example: We can work the same problem as above, but determine $\dot{V}O_{2R}$. This is computed by subtracting the resting $\dot{V}O_2$ from the maximal $\dot{V}O_2$, similar to the Karvonen equation to calculate heart rate reserve (HRR). Resting $\dot{V}O_2$ is assumed to be 3.5 mL/kg/min. The intensity is again 50% (0.50):

Step 1: Determine $\dot{V}O_2$ intensity based on $\dot{V}O_2$ reserve.

$$\dot{V}O_{2R} = ([\dot{V}O_{2max} - \dot{V}O_{2rest}] \cdot \text{intensity}) + \dot{V}O_{2rest}$$

$$= ([21.5 - 3.5] \cdot 0.50) \qquad + 3.5$$

$$= (18 \cdot 0.50) \qquad\qquad\quad + 3.5$$

$$= 9 \qquad\qquad\qquad\qquad + 3.5$$

$$= 12.50 \text{ mL/kg/min}$$

Step 2: Use the training $\dot{V}O_2$, solve the walking equation to determine speed.

$$\dot{V}O_2 \text{ (mL/kg/min)} = [S \text{ (m/min)} \cdot 0.1] + [S \text{ (m/min)} \cdot G \text{ (dec)} \cdot 1.8] + 3.5$$
$$12.5 = [S \text{ m/min} \cdot 0.10] + [S \text{ m/min} \cdot 0 \cdot 1.8] + 3.5$$
$$-3.5 \qquad\qquad\qquad\qquad\qquad -3.5$$
$$9.0 = [S \text{ m/min} \cdot 0.10] + 0$$
$$\frac{9.0}{0.10} = [S \text{ m/min} \cdot \frac{0.10}{0.10}$$
$$90 = S \text{ m/min}$$

Step 3: Convert speed in m/min to miles/hr.

$$\text{Remember: } 1 \text{ mile/hr} = 26.8 \text{ m/min}$$
$$\frac{90 \text{ m/min}}{26.8} = \textbf{3.4 mph}$$

Answer: The walking speed to achieve 50% of $\dot{V}O_{2R}$ is 3.4 miles/hr.

Calculating Walking Speed for $\dot{V}O_2$ Values When Walking Is Performed at an Incline: Using Grade

In some cases, the desired $\dot{V}O_2$ is so high that it exceeds a level walking pace. In this case, you must select a walking pace that seems appropriate, and then compute the grade necessary to elicit the desired $\dot{V}O_2$.

Example: Steve typically jogs at a pace equivalent to a $\dot{V}O_2$ of 30.0 mL/kg/min. Due to an injury, you wish to have Steve walk at 3.5 mph, yet walk on an incline that will elicit the same $\dot{V}O_2$ (30.0 mL/kg/min). What is the appropriate grade?

Step 1: Convert speed in miles per hour to meters per minute (1 mile/hr = 26.8 m/min).

$$\text{Target speed} = 3.5 \text{ miles/hr}$$
$$3.5 \cdot 26.8 = 93.8 \text{ m/min}$$

Step 2: Using the walking equation, solve for grade.

$$\dot{V}O_2 \text{ (mL/kg/min)} = [S \text{ (m/min)} \cdot 0.1] + [S \text{ (m/min)} \cdot G \text{ (dec)} \cdot 1.8] + 3.5$$
$$30.0 = [93.8 \cdot 0.10] + [93.8 \cdot G \cdot 1.8] + 3.5$$
$$30.0 = 9.38 + 168.84 \, G + 3.5$$
$$-3.5 \qquad\qquad\qquad -3.5$$
$$26.5 = 9.38 + 168.84 \, G$$
$$-9.38 \quad -9.38$$

$$\frac{17.12}{168.84} = \frac{168.84 \ G}{168.84}$$

$$0.10 = G$$

Step 3: Convert grade from a fraction (decimal) to percent grade.

$$\% \text{ grade} = \text{fraction (dec)} \cdot 100$$
$$= 0.10 \cdot 100$$
$$= \textbf{10\% Grade}$$

Answer: The grade required to elicit a $\dot{V}O_2$ of 30.0 ml/kg/min while walking 3.5 miles/hr is 10%.

Common Clinical Use of the Walking Equation: Using a MET Prescription

It is common in the clinical setting for cardiac rehab patients to report to phase II rehab with an initial MET prescription, probably based upon the results of a low-level graded exercise test. The exercise specialist will then take the MET prescription and choose initial walking and cycling intensities.

Example: Your phase II cardiac rehab patient begins his first week. The exercise plan is to begin exercising at 2.5 METs. Calculate the appropriate walking speed on level terrain (which means 0% grade).

Step 1: Convert METs to $\dot{V}O_2$ (mL/kg/min).

$$\dot{V}O_2 \text{ (mL/kg/min)} = \text{METs} \cdot 3.5$$
$$= 2.5 \cdot 3.5$$
$$= 8.75 \text{ mL/kg/min}$$

Step 2: Use the walking equation and solve for speed.

$$\dot{V}O_2 \text{ mL/kg/min} = [\text{S (m/min)} \cdot 0.1] + [\text{S (m/min)} \cdot \text{G (dec)} \cdot 1.8] + 3.5$$
$$8.75 = [\text{S m/min} \cdot 0.10] + [\text{S m/min} \cdot 0 \cdot 1.8] + 3.5$$
$$8.75 = [\text{S m/min} \cdot 0.10] + 0 + 3.5$$
$$-3.5 \qquad\qquad\qquad\qquad\qquad -3.5$$
$$\frac{5.25}{0.10} = \text{S m/min} \cdot \frac{0.10}{0.\,10}$$
$$52.5 = \text{S m/min}$$

Step 3: Convert speed in m/min to miles/hr (1 mile/hr = 26.8 m/min).

$$\frac{52.5 \text{ m/min}}{26.8} = \textbf{1.96 miles/hr}$$

PRACTICE TABLE

PRACTICE TABLE 2. *Fill in the appropriate information and complete the table (answers start on page 33).*

Problem #	$\dot{V}O_2$ mL/kg/min	METs	Walking Speed Miles/hr	% Grade
1	12.0			0
2	12.0			2
3		4.00		0
4		12.00	3.80	
5	20.0			5
6		3.00		0

ANSWERS TO PRACTICE TABLE

PRACTICE TABLE 1

Problem #	Walking speed (miles/hr)	Speed m/min	Walking grade (%)	Grade (decimal)	Body weight (kg)	$\dot{V}O_2$ mL/kg/min	$\dot{V}O_2$ mL/min	$\dot{V}O_2$ L/min	kcals/min
1	2.0	53.60	5	0.05	50	13.68	684.00	0.684	3.42
2	3.4	91.12	0	0	90	12.61	1134.90	1.135	5.68
3	3.1	83.08	8	0.08	62	23.77	1473.74	1.474	7.37
4	1.7	45.56	10	0.10	75	16.26	1219.50	1.220	6.10

SOLUTIONS (SEE TABLE 5.1 FOR CONVERSIONS)

Problem 1

$$\dot{V}O_2 \ (\text{mL/kg/min}) = [S \text{ m/min} \cdot 0.10] + [S \text{ m/min} \cdot G \cdot 1.8] + 3.5$$
$$= [53.6 \cdot 0.10] + [53.6 \cdot 0.05 \cdot 1.8] + 3.5$$
$$= 5.36 + 4.82 + 3.5$$
$$= \mathbf{13.68 \ mL/kg/min}$$

Problem 2

$$\dot{V}O_2 \ (\text{mL/kg/min}) = [S \text{ m/min} \cdot 0.10] + [S \text{ m/min} \cdot G \cdot 1.8] + 3.5$$
$$= [91.12 \cdot 0.10] + [91.12 \cdot 0 \cdot 1.8] + 3.5$$
$$= 9.11 + 0 + 3.5$$
$$= \mathbf{12.61 \ mL/kg/min}$$

Problem 3

$$\dot{V}O_2 \,(mL/kg/min) = [S \text{ m/min} \cdot 0.10] + [S \text{ m/min} \cdot G \cdot 1.8] + 3.5$$
$$= [83.08 \cdot 0.10] + [83.08 \cdot 0.08 \cdot 1.8] + 3.5$$
$$= 8.31 + 11.96 + 3.5$$
$$= \mathbf{23.77 \ mL/kg/min}$$

Problem 4

$$\dot{V}O_2 \,(mL/kg/min) = [S \text{ m/min} \cdot 0.10] + [S \text{ m/min} \cdot G \cdot 1.8] + 3.5$$
$$= [45.56 \cdot 0.10] + [45.56 \cdot 0.10 \cdot 1.8] + 3.5$$
$$= 4.56 + 8.2 + 3.5$$
$$= \mathbf{16.26 \ mL/kg/min}$$

$\dot{V}O_2$ **in mL/min—Computed by multiplying $\dot{V}O_2$ (mL/kg/min) by body weight (kg) $\dot{V}O_2$ in L/min—Computed by dividing $\dot{V}O_2$ (mL/min) by 1000 kcals/min—Computed by multiplying $\dot{V}O_2$ (L/min) by 5.**

PRACTICE TABLE 2

Problem #	$\dot{V}O_2$ mL/kg/min	METs	Walking Speed	% Grade
1	12.00	3.43	3.17	0
2	12.00	3.43	2.33	2
3	14.00	4.00	3.90	0
4	42.00	12.00	3.80	15
5	20.00	5.71	3.24	5
6	10.50	3.00	2.61	0

METs to $\dot{V}O_2$ mL/kg/min conversion: $\dot{V}O_2$ (mL/kg/min) = METs · 3.5 mL/kg/min

Miles/hr to m/min conversion: m/min $= \dfrac{\text{miles/hr}}{26.8}$

SOLUTIONS

Problem 1 (solve for walk speed)

$$\dot{V}O_2 \,(mL/kg/min) = [S \text{ m/min} \cdot 0.10] + [S \text{ m/min} \cdot G \cdot 1.8] + 3.5$$
$$12.0 = [S \text{ m/min} \cdot 0.10] + [S \text{ m/min} \cdot 0 \cdot 1.8] + 3.5$$
$$12.0 = [S \text{ m/min} \cdot 0.10] + 0 + 3.5$$
$$-3.5 \qquad\qquad\qquad\qquad\qquad -3.5$$
$$\frac{8.5}{0.10} = S \text{ m/min} \cdot \frac{0.10}{0.10}$$

$$85 = S \text{ m/min}$$

$$\frac{85}{26.8} = 3.17 \text{ miles/hr}$$

Problem 2 (solve for walk speed)

$$\dot{V}O_2 \text{ (mL/kg/min)} = [S \text{ m/min} \cdot 0.10] + [S \text{ m/min} \cdot G \cdot 1.8] + 3.5$$

$$12.0 = [S \text{ m/min} \cdot 0.10] + [S \text{ m/min} \cdot 0.02 \cdot 1.8] + 3.5$$

$$12.0 = S \text{ m/min } 0.10 + S \text{ m/min } 0.036 + 3.5$$

$$-3.5 \qquad\qquad\qquad -3.5$$

$$\frac{8.5}{0.136} = S \text{ m/min} \cdot \frac{0.136}{0.136}$$

$$62.5 = \text{m/min}$$

$$\frac{62.5}{26.8} = 2.33 \text{ miles/hr}$$

Problem 3 (solve for walk speed)

$$\dot{V}O_2 \text{ (mL/kg/min)} = [S \text{ m/min} \cdot 0.10] + [S \text{ m/min} \cdot G \cdot 1.8] + 3.5$$

$$14.0 = [S \text{ m/min} \cdot 0.10] + [S \text{ m/min} \cdot 0 \cdot 1.8] + 3.5$$

$$14.0 = [S \text{ m/min} \cdot 0.10] + 0 + 3.5$$

$$-3.5 \qquad\qquad\qquad -3.5$$

$$\frac{10.5}{0.10} = S \text{ m/min} \cdot \frac{0.10}{0.10}$$

$$105 = S \text{ m/min}$$

$$\frac{105}{26.8} = 3.9 \text{ miles/hr}$$

Problem 4 (solve for grade)

Walk speed = 3.8 miles/hr   m/min = miles/hr · 26.8   so, 3.8 · 26.8 = 101.8 m/min

$$\dot{V}O_2 \text{ (mL/kg/min)} = [S \text{ m/min} \cdot 0.10] + [S \text{ m/min} \cdot G \cdot 1.8] + 3.5$$

$$42.0 = [101.8 \cdot 0.10] + [101.8 \cdot G \cdot 1.8] + 3.5$$

$$42.0 = 10.2 + 183.2 \, G + 3.5$$

$$-3.5 \qquad\qquad\qquad -3.5$$

$$38.5 = 10.2 + 183.2 \, G$$

$$-10.2 \quad -10.2$$

$$\frac{28.3}{183.2} = \frac{183.2 \, G}{183.2}$$

$$0.15 = G$$

Convert grade (decimal) to percent: = Grade (decimal) · 100
$$= 0.15 \cdot 100 = 15\%$$

Problem 5 (solve for walk speed)

$$\dot{V}O_2 \text{ (mL/kg/min)} = [\text{S m/min} \cdot 0.10] + [\text{S m/min} \cdot G \cdot 1.8] + 3.5$$
$$20.0 = [\text{S m/min} \cdot 0.10] + [\text{S m/min} \cdot 0.05 \cdot 1.8] + 3.5$$
$$20.0 = 0.10 \text{ m/min} + 0.09 \text{ m/min} + 3.5$$
$$-3.5 \qquad\qquad\qquad -3.5$$
$$\frac{16.5}{0.19} = \frac{0.19 \text{ m/min}}{0.19}$$
$$86.8 = \text{m/min}$$
$$\frac{86.8}{26.8} = \textbf{3.24 miles/hr}$$

Problem 6 (solve for walk speed)

$$\dot{V}O_2 \text{ (mL/kg/min)} = [\text{S m/min} \cdot 0.10] + [\text{S m/min} \cdot G \cdot 1.8] + 3.5$$
$$10.5 = [\text{S m/min} \cdot 0.10] + [\text{S m/min} \cdot 0 \cdot 1.8] + 3.5$$
$$10.5 = [\text{S m/min} \cdot 0.10] + 0 + 3.5$$
$$-3.5 \qquad\qquad\qquad -3.5$$
$$\frac{7.0}{0.10} = \text{S m/min} \cdot \frac{0.10}{0.10}$$
$$70 = \text{m/min}$$
$$\frac{70}{26.8} = \textbf{2.61 miles/hr}$$

TABLE 5.1 **Essential Conversions**

Action	Conversion
Conversion of miles per hour to meters per minute	Miles/Hr $\cdot$ 26.8 = m/min
Conversion of percent grade to grade as fraction	$\dfrac{\% \text{ Grade}}{100}$
Conversion of $\dot{V}O_2$ in mL/kg/min to METs	$\dfrac{\dot{V}O_2 \text{ (mL/kg/min)}}{3.5}$
Conversion of $\dot{V}O_2$ in mL/kg/min to mL/min	$\dot{V}O_2$ (mL/kg/min) $\cdot$ body wt. in KG
Conversion of $\dot{V}O_2$ in mL/min to L/min	$\dfrac{\dot{V}O_2 \text{ (mL/min)}}{1000}$
Conversion of $\dot{V}O_2$ in L/min to kilocalories/min	$\dot{V}O_2$ L/min $\cdot$ 5

RUNNING EQUATION

DERIVATION OF THE RUNNING EQUATION

As in the case with the walking equation, the running equation is composed of three components: a horizontal, vertical, and resting component.

The horizontal component of the running equation requires twice the oxygen compared to horizontal walking. Thus, the horizontal component is computed as:

Horizontal Component = S (m/min) × 0.2 mL/kg/m

The vertical component does not have the same oxygen cost for running as it does for walking. The vertical component can be computed as:

Vertical Component = S (m/min) × G (frac) × 0.9 mL/min/m

Note that grade is expressed as a fraction or decimal. Thus a 10% grade is expressed as 0.10 and a 5% grade is expressed as 0.05.

The resting component is the same for the running equation as the walking equation, and is equivalent to 1 MET or 3.5 mL/kg/min.

Resting Component = 3.5 mL/kg/min

ACSM RUNNING EQUATION

The ACSM running equation is:

$$\dot{V}O_2 \text{ (mL/kg/min)} = (S \text{ m/min} \times 0.2) + (S \text{ m/min} \times G \times 0.9) + 3.5$$

Limitations of the Running Equation

Speed Range for Running Equation Accuracy: Generally, the running equation is designed for speeds greater than 5.0 mph (134 m/min). However, depending on body size and preference, some individuals may choose to run at speeds as low as 3.7 mph (100 m/min). Individual analysis will determine which equation (walk or run) should be used. For example, if the client is jogging at 4.0 mph, then it would be appropriate to use the running equation.

Steady State Exercise: As with the walking equation, the running equation is only valid with steady state exercise.

SPECIFIC USES OF THE RUNNING EQUATION

Example 1: Solve for $\dot{V}O_2$ and Calculate Calorie Costs

An 80 kg individual is jogging 7.0 mph up a 2% incline. What is the oxygen cost? How many kilocalories are expended?

Step 1: Convert speed and grade: (1 m/hr = 26.8 m/min).

$$7.0 \text{ mile/hr} \times 26.8 = \textbf{187.6 m/min}$$

$$\frac{2\% \text{ grade}}{100} = \textbf{0.02}$$

Step 2: Put in values and solve.

$$\dot{V}O_2 \text{ (mL/kg/min)} = (\text{S m/min} \cdot 0.2) + (\text{S m/min} \cdot \text{grade} \cdot 0.9) + 3.5$$
$$= (187.6 \cdot 0.2) \qquad + (187.6 \cdot 0.02 \cdot 0.9) \qquad + 3.5$$
$$= 37.52 \qquad\qquad + 3.377 \qquad\qquad\qquad + 3.5$$
$$= \textbf{44.4 mL/kg/min}$$

Step 3: Convert $\dot{V}O_2$ (mL/kg/min) to L/min

$$44.4 \text{ mL/kg/min} \cdot 80 \text{ kg} = 3552 \text{ mL/min}$$

$$\frac{3552 \text{ mL/min}}{1000} = \textbf{3.55 L/min}$$

Step 4: Convert $\dot{V}O_2$ L/min to kcal/min.

$$\text{Computed as kcals} = \dot{V}O_2/\text{L/min} \cdot 5$$
$$= 3.55 \cdot 5 = \textbf{17.75 kcals/min}$$

Example 2: Outdoor Running on a Flat Surface

In many cases with outdoor running, running pace is computed as a "pace" from minutes per mile. Converting a running pace from minutes per mile into miles per hour is performed by dividing 60 min/hr by the pace in min/mile.

A 70 kg individual is running at a 7:30/mile pace (7 min 30 sec). Assume he runs over level ground. He does this for 30 minutes. How many total calories are expended?

Step 1: Convert run pace (min/mile) into miles/hr.
Convert 30 seconds into a fraction (decimal).

$$\frac{30 \text{ seconds/min}}{60 \text{ seconds/min}} = 0.5 \text{ seconds/min}$$

Now convert running pace to a speed.

$$\frac{60 \text{ min/hr}}{7.5 \text{ min/mile}} = \textbf{8.0 miles/hr}$$

Step 2: Convert miles/hr into m/min.

$$8 \text{ miles/hr} \cdot 26.8 = \textbf{214.4 m/min}$$

Step 3: Using equation, solve for $\dot{V}O_2$. Note: Grade is given as 0%.

$$\dot{V}O_2 \text{ mL/kg/min} = (214.4 \cdot 0.2) + (214.4 \cdot G \cdot 0.9) + 3.5$$
$$= 42.88 + 0 + 3.5$$
$$= \textbf{46.38 mL/kg/min}$$

Step 4: Convert $\dot{V}O_2$ mL/kg/min to L/min.

$$46.3 \text{ mL/kg/min} \cdot 70\text{kg} = 3246.6 \text{ mL/min}$$

$$\frac{3236.6}{1000} = \textbf{3.25 L/min}$$

Step 5: Convert $\dot{V}O_2$ L/min to kcals/min.

$$3.25 \cdot 5 = 16.25 \text{ kcals/min} \cdot 30 \text{ minutes of exercise} = \textbf{487.5 kcals}$$

Example 3: Exercise Prescription

A common use of the running formula is to determine an appropriate running speed based upon a $\dot{V}O_2$ value.

An individual has their $\dot{V}O_{2max}$ measured at 52.0 mL/kg/min. You decide on a training intensity equivalent to 75% of $\dot{V}O_{2max}$. What is the running speed (assume 0% grade)?

Step 1: Determine 75% of $\dot{V}O_{2max}$.

$$52 \text{ mL/kg/min} \cdot 0.75 = \textbf{39.0 mL/kg/min}$$

Step 2: Use running equation (assume 0% grade) and solve for speed.

$$\dot{V}O_2\text{(mL/kg/min)} = (S\text{ m/min} \times 0.2) + (S\text{ m/min} \times G \times 0.9) + 3.5$$
$$39\text{ mL/kg/min} = (S\text{ m/min} \times 0.2) + 0 + 3.5$$
$$-3.5 \qquad\qquad\qquad\qquad\qquad\qquad -3.5$$
$$\frac{35.5}{0.2} = S\text{m/min} \times \frac{0.2}{0.2}$$
$$\mathbf{177.5 = S\ m/min}$$

Step 3: Convert m/min to miles/hr.

$$\frac{177.5}{26.8} = \textbf{6.6 miles/hr}$$

Example 4

Convert a walk-plus-grade work rate into a level running speed.

The Bruce protocol, a common clinical exercise testing protocol was used to assess the fitness of a young, healthy adult. You determine that based on a perception of effort, stage 3 (3.4 mph 14% grade) seems to be the appropriate intensity. What would the equivalent intensity be in running speed (0% grade)?

Step 1: Determine $\dot{V}O_2$ for Stage 3 of the Bruce protocol (3.4 miles/hr 14% grade).

Convert speed to m/min: $3.4 \cdot 26.8 = \textbf{91.12 m/min}$
Convert grade to decimal: $14/100 = \textbf{0.14}$

Solve for $\dot{V}O_2$ using the *walking* equation.

$$\dot{V}O_2\text{(mL/kg/min)} = (S\text{ m/min} \cdot 0.1) + (S\text{ m/min} \cdot G \cdot 1.8) + 3.5$$
$$= (91.12 \cdot 0.1) + (91.12 \cdot 0.14 \cdot 1.8) + 3.5$$
$$= 9.11 + 22.96 + 3.5$$
$$= \textbf{35.57 mL/kg/min}$$

Step 3: Using the $\dot{V}O_2$ calculated from Step 2, use the running equation to determine running speed on level terrain (0% grade).

$$\dot{V}O_2\text{(mL/kg/min)} = (S\text{ m/min} \cdot 0.2) + (S\text{ m/min} \cdot G \cdot 0.9) + 3.5$$
$$35.57 = (S\text{ m/min} \cdot 0.2) + 0 + 3.5$$
$$-3.5 \qquad\qquad\qquad\qquad\qquad -3.5$$
$$30.57 = S\text{ m/min} \cdot 0.2$$
$$\frac{30.57}{0.2} = S\text{ m/min} \cdot \frac{0.2}{0.2}$$
$$\mathbf{152.9 = S\ m/min}$$

Step 4: Convert m/min to mph

$$\frac{152.9}{26.8} = \textbf{5.7 miles/hr}$$

PRACTICE TABLE

PRACTICE TABLE 1. *Fill in the appropriate information and complete the table (answers start below).*

Problem	$\dot{V}O_2$ mL/kg/	METs	Body wt. (kg)	kcals/min	Speed (mph)	Grade (%)
1			60		6.5	2
2	42		85			0
3		18	52			2
4			90	14		0
5			70		9	0
6		15	75			0

SOLUTIONS

Problem 1

Compute $\dot{V}O_2$ in mL/kg/min:

Speed conversion: 6.5 mph · 26.8 = 174.2 m/min

Grade Conversion: $= \dfrac{2\% \text{ grade}}{100} = 0.02$

$$\begin{aligned}
\dot{V}O_2/\text{mL/kg/min} &= (\text{S m/min} \cdot 0.2) + (\text{S m/min} \cdot \text{G} \cdot 0.9) + 3.5 \\
&= (174.2 \cdot 0.2) + (174.2 \cdot 0.02 \cdot 0.9) + 3.5 \\
&= 34.84 + 3.13 + 3.5
\end{aligned}$$

$$\dot{V}O_2 \text{ (mL/kg/min)} = \textbf{41.5}$$

Convert $\dot{V}O_2$ in mL/kg/min into METs:

$$\text{METs} = \frac{\dot{V}O_2 \text{(mL/kg/min)}}{3.5}$$

$$\frac{41.5}{3.5} = \textbf{11.9 MET}$$

KCAL/MIN DETERMINATION:

Convert $\dot{V}O_2$ (mL/kg/min) into L/min:

$$\dot{V}O_2 \text{ (mL/kg/min)} \cdot \text{body wt. (kg)}$$
$$41.5 \cdot 60 = 2490 \text{ mL/min}$$

Convert to mL/min to L/min (divide by 1000):

$$\frac{2490 \text{ mL/min}}{1000} = 2.49 \text{ L/min}$$

ANSWERS TO PRACTICE TABLE

PRACTICE TABLE 1.

Problem	$\dot{V}O_2$ mL/kg/min	METs	Body wt. (kg)	kcals/min	Speed (miles)/h	Grade (%)
1	41.5	11.9	60	12.45	6.5	2
2	42	12	85	17.85	7.2	0
3	63	18	52	16.5	10.5	2
4	31.1	8.9	90	14	5.1	0
5	51.7	14.8	70	18	9	0
6	52.5	15	75	19.5	9.1	0

Convert $\dot{V}O_2$ L/min to kcals/min (L/min · 5):

$$2.49 \text{ L/min} \cdot 5 = \textbf{12.45 kcals/min}$$

Problem 2

Convert $\dot{V}O_2$ in mL/kg/min into METs:

$$\text{METs} = \frac{\dot{V}O_2(\text{mL/kg/min})}{3.5}$$

$$\frac{42.0}{3.5} = \textbf{12 METs}$$

KCAL/MIN DETERMINATION:
Convert $\dot{V}O_2(\text{mL/kg/min})$ into L/min:

$$\dot{V}O_2 \text{ (mL/kg/min} \cdot \text{body wt. (kg)}$$
$$42 \cdot 85 = 3570 \text{ mL/min}$$

Convert to mL/min to L/min (divide by 1000):

$$\frac{3570 \text{ mL/min}}{1000} = 3.57 \text{ L/min}$$

Convert $\dot{V}O_2$ L/min to kcals min (L/min · 5):

$$3.57 \text{ L/min} \cdot 5 = \textbf{17.85 kcals/min}$$

Use running equation to solve for speed:

$$\dot{V}O_2 \text{ (mL/kg/min)} = (\text{S m/min} \cdot 0.2) + (\text{S m/min} \cdot \text{G} \cdot 0.9) + 3.5$$
$$42.0 = (\text{m/min} \cdot 0.2) + (\text{m/min} \cdot 0 \cdot 0.9) + 3.5$$
$$-3.5 \qquad\qquad\qquad\qquad\qquad\qquad\qquad\qquad -3.5$$

$$\frac{38.5}{0.2} = \left(S \text{ m/min} \cdot \frac{0.2}{0.2} \right) + 0$$

$$192.5 = S \text{ m/min}$$

Convert m/min to miles/hr (m/min divided by 26.8):

$$\frac{192.5}{26.8} = \textbf{7.2 miles/hr}$$

Problem 3

Convert $\dot{V}O_2$ in METs to $\dot{V}O_2$ in mL/kg/min:

$$METs \cdot 3.5$$

$$18 \cdot 3.5 = \textbf{63 mL/kg/min}$$

KCAL/MIN DETERMINATION:

Convert $\dot{V}O_2$ (mL/kg/min) into L/min.

$$\dot{V}O_2 \text{ (mL/kg/min)} \times \text{body wt. (kg)}$$

$$63 \cdot 52 = 3276 \text{ mL/min}$$

Convert to mL/min to L/min (divide by 1000):

$$\frac{3276 \text{ mL/min}}{1000} = 3.3 \text{ L/min}$$

Convert $\dot{V}O_2$ L/min to kcals min (L/min · 5):

$$3.3 \text{ L/min} \cdot 5 = \textbf{16.5 kcals/min}$$

Use running equation to solve for speed (note 2% grade = 0.02):

$$\dot{V}O_2 \text{ (mL/kg/min)} = (S \text{ m/min} \times 0.2) + (S \text{ m/min} \times G \times 0.9) + 3.5$$

$$63.0 = (S \text{ m/min} \times 0.2) + (S \text{ m/min} \times 0.02 \times 0.9) + 3.5$$

$$-3.5 \qquad\qquad\qquad\qquad\qquad\qquad\qquad -3.5$$

$$59.5 = 0.2 \text{ S m/min} + 0.018 \text{ S m/min}$$

$$59.5 = \frac{0.218}{0.218} \text{ m/min}$$

$$272.9 = S \text{ m/min}$$

Convert m/min to miles/hr (m/min divided by 26.8):

$$\frac{272.9}{26.8} = \textbf{10.2 miles/hr}$$

Problem 4

Convert kcals/min to $\dot{V}O_2$ in mL/kg/min:

$$\frac{\text{kcals/min}}{5} = \dot{V}O_2 \text{ L/min}$$

$$\frac{14}{5} = 2.8 \text{ L/min}$$

$$2.8 \text{ L/min} \cdot 1000 = 2800 \text{ mL/min}$$

Divide by body weight to determine $\dot{V}O_2$ mL/kg/min:

$$\frac{2800}{90} = \textbf{31.1 mL/kg/min}$$

Convert $\dot{V}O_2$ mL/kg/min to METs:

$$\frac{31.1}{3.5} = \textbf{8.9 METs}$$

Use running equation to solve for speed (note 0% grade):

$$\dot{V}O_2 \text{ (mL/kg/min)} = (\text{S m/min} \times 0.2) + (\text{S m/min} \times \text{G} \times 0.9) + 3.5$$
$$31.1 = (\text{S m/min} \times 0.2) + (\text{S m/min} \times 0 \times 0.9) + 3.5$$
$$-3.5 \qquad\qquad\qquad\qquad\qquad\qquad\qquad\qquad -3.5$$
$$\frac{27.6}{0.2} = \text{S m/min} \times \frac{0.2}{0.2} + 0$$
$$138 = \text{S m/min}$$

Convert m/min to mph (m/min divided by 26.8):

$$\frac{138}{26.8} = \textbf{5.1 miles/hr}$$

Problem 5

Determine $\dot{V}O_2$ in mL/kg/min (note: speed conversion 9.0 mph $\cdot$ 26.8 = 241.2 m/min):

$$\dot{V}O_2/\text{mL/kg/min} = (\text{S m/min} \times 0.2) + (\text{S m/min} \times \text{G} \times 0.9) + 3.5$$
$$= (241.2 \times 0.2) + (\text{S m/min} \times 0 \times 0.9) + 3.5$$
$$= 48.2 + 0 + 3.5$$
$$= \textbf{51.7 mL/kg/min}$$

Convert $\dot{V}O_2$ mL/kg/min into METs:

$$\frac{\dot{V}O_2 \text{ (mL/kg/min)}}{3.5} = \text{METs}$$

$$\frac{51.7}{3.5} = \textbf{14.8 METs}$$

KCAL/MIN DETERMINATION:

Convert $\dot{V}O_2$ mL/kg/min into L/min:

$$\dot{V}O_2 \text{ (mL/kg/min)} \cdot \text{body wt. (kg)}$$
$$51.7 \cdot 70 = 3619 \text{ mL/min}$$

Convert to mL/min to L/min (divide by 1000):

$$\frac{3619 \text{ mL/min}}{1000} = 3.6 \text{ L/min}$$

Convert $\dot{V}O_2$ L/min to kcals min (L/min · 5):

$$3.6 \text{ L/min} \cdot 5 = \textbf{18 kcals/min}$$

Problem 6

Convert METs into $\dot{V}O_2$ in mL/kg/min:

$$\text{METs} \times 3.5 = \dot{V}O_2 \text{ (mL/kg/min)}$$
$$15 \times 3.5 = \textbf{52.5 mL/kg/min}$$

Use running equation to solve for speed (note 0% grade):

$$\dot{V}O_2 \text{ (mL/kg/min)} = (\text{S m/min} \times 0.2) + (\text{S m/min} \times G \times 0.9) + 3.5$$
$$52.5 = (\text{S m/min} \times 0.2) + (\text{S m/min} \times 0 \times 0.9) + 3.5$$
$$-3.5 \qquad\qquad\qquad\qquad\qquad\qquad\qquad\qquad -3.5$$
$$\frac{49.0}{0.2} = \text{S m/min} \cdot \frac{0.2}{0.2} + 0$$
$$245 = \text{S m/min}$$

Convert m/min to miles/hr (m/min divided by 26.8):

$$\frac{245}{26.8} = \textbf{9.1 miles/hr}$$

KCAL/MIN DETERMINATION:

Convert $\dot{V}O_2$ (mL/kg/min) into L/min:

$$\dot{V}O_2 \text{ (mL/kg/min)} \cdot \text{body wt. (kg)}$$

$$52.5 \cdot 75 = 3937.5 \text{ mL/min}$$

Convert to mL/min to L/min (divide by 1000):

$$\frac{3937.5}{1000} = 3.9 \text{ L/min}$$

Convert $\dot{V}O_2$ L/min to kcals min (L/min · 5):

$$3.9 \text{ L/min} \cdot 5 = \textbf{19.5 kcals/min}$$

LEG ERGOMETER EQUATION

The leg ergometer, or cycle ergometer as it is commonly called, is one of the most common modes of both testing and exercise. Leg ergometers are commonly used for submaximal exercise testing, and aerobic training in the health-fitness setting.

DERIVATION OF THE LEG ERGOMETER EQUATION

During leg ergometer exercise, the client is pedaling against a resistance, which moves a flywheel a given distance with each turn of the pedal crank. Computing the number of pedal turns n a minute helps us compute the work rate. Work rate during leg ergometer exercise is computed in kilogram · meters per minute (kg · m/min):

$$\text{work rate (kg} \cdot \text{m/min)} = R \text{ (kg)} \cdot D \text{ (m)} \cdot \text{Rev/min}$$

Where R = Resistance or load, expressed as kilograms
    D = Distance in meters the flywheel travels with each pedal crank
Rev/min = Revolutions (pedal cranks) per minute

A Note on D

The flywheel on a leg ergometer may vary by manufacturer. Therefore, depending on the ergometer manufacturer, the distance the flywheel travls with each pedal revolution will vary:

Monark leg ergometers: 6 m per pedal revolution
Tunturi leg ergometers: 3 m per revolution
Monark arm ergometers: 2.4 m per revolution

Be sure to identify the type of leg ergometer being used so that you can accuately quantify work rate.

A Note on Work Rate

Another way to express work rate is in watts. These units are commonly used in health and fitness settings as well as in the research literature. Conversion from $kg \cdot m/min$ to watts is quite simple:

$$\text{watts} = \frac{kg \cdot m/min}{6.12} \qquad \text{or} \qquad kg \cdot m/min = \text{watts} \times 6.12$$

OXYGEN COST OF LEG ERGOMETER EXERCISE

Leg ergometer exercise involves three components when calculating the oxygen cost $(\dot{V}O_2)$ of work.

1. Oxygen cost of unloaded cycling. The oxygen cost of simply moving the flywheel as well as the movement of the legs themselves is approximately 3.5 mL/kg/min. This applies to pedal revolutions of 50–60 rev/min, so the cost may vary when pedaling outside of this range.
2. External resistance or load placed on the flywheel, which is approximately 1.8 mL/kg/min for each $kg \cdot m/min$.
3. Resting oxygen consumption, which is approximately 3.5 mL/kg/min.

THE ACSM LEG ERGOMETER EQUATION

$$\dot{V}O_2 \ mL \cdot kg^{-1} \cdot min^{-1} = \left[1.8 \ \frac{\text{work rate } (kg \cdot m/min)}{\text{body mass } (kg)} \right]$$
$$+ \ (3.5 \ mL/kg/min) + (3.5 \ mL/kg/min)$$

Expressed more simply it is:

$$\dot{V}O_2 \ (mL \cdot kg^{-1} \cdot min^{-1}) = 1.8 \ \frac{\text{work rate } (kg \cdot m/min)}{\text{body mass } (kg)} + 7$$

Limitations of the Leg Ergometer Equation

1. *Steady-State Exercise Only!* The ACSM leg ergometer equation was computed based upon steady-state oxygen consumption values. Therefore, the equation will not be as accurate when computing oxyen consumption during non–steady-state exercise. Using the ACSM leg ergometer equation to compute a maximal oxygen uptake (for example, the last stage of a maximal exercise test) will not be accurate during such non–steady-state exercise.
2. Work rates fom 300–1200 $kg \cdot m/min$: In watts, the accuracy is between approximately 50 and 200. Fit individuals may exercise beyond these work rates, so the potential error increases as work rates exceed this range.

SPECIFIC USES OF THE LEG ERGOMETER EQUATION
Determine the Caloric Cost of Leg Ergometer Exercise

When an individual is cycling, you can compute first their oxygen cost ($\dot{V}O_2$) and then determine their caloric cost.

Example: An individual is cycling at a load (R) of 1.5 kg on a Monark cycle, pedaling 60 rev/min and weighing 75 kg.

Step 1: Determine work rate in kg · m/min:

$$\text{work rate (kgm} \cdot \text{min}^{-1}) = R \text{ (kg)} \cdot D \text{ (m)} \cdot \text{rev/min}$$

R = 1.5 kg   D = 6 m (because it is a Monark cycle)   rev/min = 60

$$\text{work rate (kgm} \cdot \text{min}^{-1}) = 1.5 \cdot 6 \cdot 60$$
$$= 540 \text{ kg} \cdot \text{m/min}$$

Step 2: Calculate $\dot{V}O_2$ (remember that body mass (BM) is 75 kg).

$$\dot{V}O_2 \text{ (mL/kg/min)} = 1.8 \frac{\text{work rate (kg} \cdot \text{m/min)}}{\text{body mass (kg)}} + 7$$

$$= 1.8 \frac{540}{75} + 7$$
$$= 1.8 \, (7.2) + 7$$
$$= 12.96 + 7$$
$$= 19.96 \text{ mL/kg/min}$$

Step 3: Convert $\dot{V}O_2$ in mL/kg/min into kcals/min.

$$19.96 \text{ mL/kg/min} \cdot 75 \text{ kg} = 1497 \text{ mL/min}$$
$$\frac{1497}{1000} = 1.5 \text{ L/min}$$
$$1.5 \cdot 5 = 7.5 \text{ kcals/min}$$

Determine the Appropriate Work Rate for an Exercise Prescription

In many cases, you have a training $\dot{V}O_2$, and need to compute work rate in kg · m/min or perhaps even watts.

Example: An 80 kg man has a training $\dot{V}O_2$ of 21.5 mL/kg/min. Compute the work rate at which he should train in both watts and kg · m/min.

Step 1: Determine kg · m/min.

$$\dot{V}O_2 \text{ (mL/kg/min)} = 1.8 \frac{\text{work rate (kg} \cdot \text{m/min)}}{\text{body mass (kg)}} + 7$$

$$21.5 = 1.8 \, \frac{\text{work rate (kg} \cdot \text{m/min)}}{80} + 7$$

$$-7 \hspace{7cm} -7$$

$$14.5 = 1.8 \, \frac{\text{work rate (kg} \cdot \text{m/min)}}{80}$$

$$\cdot \, 80 \hspace{7cm} \cdot \, 80$$

$$1160 = 1.8 \, (\text{kg} \cdot \text{m/min})$$

$$\frac{1160}{1.8} = \frac{1.8}{1.8} \, \text{kg} \cdot \text{m/min}$$

$$644 = \text{kg} \cdot \text{m/min}$$

Step 2: Convert to watts.

$$\text{watts} = \frac{\text{kg} \cdot \text{m/min}}{6.12}$$

$$\text{watts} = \frac{644}{6.12}$$

$$\text{watts} = 105$$

Estimate $\dot{V}O_{2max}$ Based on a Submaximal Exercise Test

In some instances, a submaximal exercise test is used to estimate $\dot{V}O_{2max}$. In this case, the equation is somewhat misapplied, since $\dot{V}O_{2max}$ would not represet steady-state exercise. However, $\dot{V}O_{2max}$ estimates are still useful to help stratify fitness and help determine appropriate exercise training intensity.

Example: The following data is for a 40-year-old client who weighs 73 kg:

Stage	Work rate (kg · m/min)	Heart rate (beats/min)
1	150	90
2	300	108
3	450	115
4	600	130

In this example, we estimate maximal work rate by extrapolating the last two work rate/heat rate plots up until we hit his age-predicted maximal heart rate (220-age) of 180. We then identify the work rate associated with the maximal heart rate, and compute $\dot{V}O_2$:

$$\dot{V}O_2 \, (\text{mL/kg/min}) = 1.8 \, \frac{\text{work rate (kg} \cdot \text{m/min)}}{\text{body mass (kg)}} + 7$$

$$= 1.8 \left(\frac{900}{73} \right) + 7$$

$$= 1.8 \, (12.3) + 7$$

$$= 22.14 + 7$$

$$= 29.14 \, \text{mL/kg/min}$$

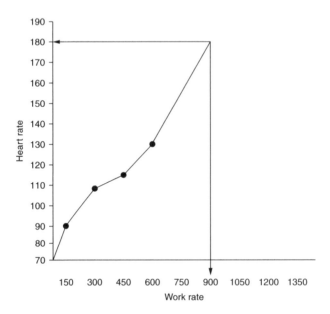

PRACTICE TABLE

PRACTICE TABLE 1. *Fill in the appropriate information and complete the table (answers start on page 50). Fill in the blanks using the following equations:*

$$\dot{V}O_2 \text{ (mL/kg/min)} = 1.8 \frac{\text{work rate (kg} \cdot \text{m/min)}}{\text{body mass (kg)}} + 7$$

$$\text{work rate (kg} \cdot \text{m/min)} = R \text{ (kg)} \times D \text{ (m)} \times RPM$$

$$\text{watts} = \frac{\text{kg} \cdot \text{m/min}}{6.12} \qquad \text{or} \qquad \text{kg} \cdot \text{m/min} = \text{watts} \cdot 6.12$$

$$\text{kcals/min} = \dot{V}O_2 \text{ L/min} \cdot 5 \qquad \text{METs} = \frac{\dot{V}O_2 \text{ mL/kg/min}}{3.5}$$

Problem	$\dot{V}O_2$ (mL/kg/min)	METs	Body wt. (kg)	kcals/ min	R (kg)	Rev/ min	Watts	Kg · m/ min
1			65		2.0	50		
2			75		1.5	50		
3			62	10	—	—		
4	24		95			50		
5			70			65	100	
6			80			60		700

Note: Assumes a Monark cycle ergometer, 6 m/rev.

PRACTICE TABLE 2. *Fill in the appropriate information and complete the table (answers start on page 57).*

Problem	$\dot{V}O_2$ (mL/kg/min)	Body mass (kg)	Kg · m/min	Watts	Running speed (no grade) miles/hr
1	35.5	70			
2	41.5	75			
3	29.2	65			
Problem	$\dot{V}O_2$ (mL/kg/min)	Body mass (kg)	Kg · m/min	Watts	Walking speed (no grade) miles/hr
1	12	70			
2	10	75			
3	7.0	65			

ANSWERS TO PRACTICE TABLES

PRACTICE TABLE 1

Problem	$\dot{V}O_2$ (mL/kg/min)	METs	Body wt. (kg)	kcals/ min	R (kg)	Rev/ min	Watts	Kg · m/ min
1	23.6	6.7	65	7.5	2.0	50	98	600
2	17.8	5.1	75	6.5	1.5	50	73.5	450
3	32.3	9.2	62	10	—	—	142	870
4	24	6.9	95	11.4	3.0	50	147	897
5	22.7	6.5	70	8.0	1.6	65	100	612
6	22.8	6.5	80	9.0	1.9	60	114	700

Note: Assumes a Monark cycle ergometer, 6 m/rev.

SOLUTIONS

Problem 1

Determine work rate:

$$\text{work rate (kg} \cdot \text{m/min)} = \text{R (kg)} \cdot \text{D (m)} \cdot \text{rev/min}$$

$$= 2.0 \cdot 6 \cdot 50$$

$$= 600 \text{ kg} \cdot \text{m/min}$$

$$\text{watts} = \frac{\text{kg} \cdot \text{m/min}}{6.12}$$

$$\frac{600}{6.12} = 98 \text{ watts}$$

Determine $\dot{V}O_2$:

$$\dot{V}O_2 \text{ (mL/kg/min)} = 1.8 \frac{\text{work rate (kg} \cdot \text{m/min)}}{\text{body mass (kg)}} + 7$$

$$= 1.8 \frac{600}{65} + 7$$

$$= 1.8 \times 9.2 + 7$$

$$= 16.56 + 7$$

$$= 23.56 \text{ mL/kg/min}$$

$$\text{METs} = \frac{\dot{V}O_2 \text{ mL/kg/min}}{3.5}$$

$$= \frac{23.56}{3.5}$$

$$= 6.7 \text{ METs}$$

Determine kcals/min (must convert $\dot{V}O_2$ into L/min):

$$23.56 \text{ mL/kg/min} \cdot 65 \text{ kg}$$

$$= 1531.4 \text{ mL/min}$$

$$= \frac{1531.4}{1000}$$

$$= 1.5 \text{ L/min}$$

$$\text{kcals/min} = \dot{V}O_2 \text{ L/min} \cdot 5$$

$$= 1.5 \cdot 5$$

$$= 7.5 \text{ kcals/min}$$

Problem 2

Determine work rate:

$$\text{work rate (kg} \cdot \text{m/min)} = \text{R (kg)} \times \text{D (m)} \times \text{rev/min}$$

$$= 1.5 \cdot 6 \cdot 50$$

$$= 450 \text{ kg} \cdot \text{m/min}$$

$$\text{watts} = \frac{\text{kg} \cdot \text{m/min}}{6.12}$$

$$\frac{450}{6.12} = 73.5 \text{ Watts}$$

Determine $\dot{V}O_2$:

$$\dot{V}O_2 \text{ (mL/kg/min)} = 1.8\,\frac{\text{work rate (kg} \cdot \text{m/min)}}{\text{body mass (kg)}} + 7$$

$$= 1.8\,\frac{450}{75} + 7$$

$$= 1.8\,(6.0) + 7$$

$$= 10.8 + 7$$

$$= 17.8 \text{ mL/kg/min}$$

$$\text{METs} = \frac{\dot{V}O_2 \text{ mL/kg/min}}{3.5}$$

$$= \frac{17.8}{3.5}$$

$$= 5.1 \text{ METs}$$

Determine kcals/min (must convert $\dot{V}O_2$ into L/min):

$$17.8 \text{ mL/kg/min} \cdot 75\text{g}$$

$$= 1335 \text{ mL/min}$$

$$= \frac{1335}{1000}$$

$$= 1.3 \text{ L/min}$$

$$\text{kcals/min} = \dot{V}O_2 \text{ L/min} \cdot 5$$

$$= 1.3 \cdot 5$$

$$= 6.5 \text{ kcals/min}$$

Problem 3

Determine $\dot{V}O_2$ from kcals/min:

$$\text{kcals/min} = \dot{V}O_2 \text{ L/min} \cdot 5$$

$$10 = \dot{V}O_2 \text{ L/min} \cdot 5$$

$$\frac{10}{5} \qquad \frac{5}{5}$$

$$2.0 = \dot{V}O_2 \text{ L/min}$$

$$2.0 \cdot 1000 = 2000 \text{ mL/min}$$

$$\frac{2000}{62 \text{ kg}} = 32.25 \text{ mL/kg/min}$$

$$\text{METs} = \frac{\dot{V}O_2 \ mL/kg/min}{3.5}$$

$$= \frac{32.25}{3.5}$$

$$= 9.2 \ \text{METs}$$

Determine kg · m/min based on $\dot{V}O_2$:

$$\dot{V}O_2 \ (mL/kg/min) = 1.8 \ \frac{\text{work rate (kg · m/min)}}{\text{body mass (kg)}} + 7$$

$$32.25 = 1.8 \ \frac{\text{kg · m/min}}{62} + 7$$

$$-7 \qquad\qquad\qquad -7$$

$$25.25 = 1.8 \ \frac{\text{kg · m/min}}{62}$$

$$\cdot \ 62 = \cdot \ 62$$

$$1565.5 = 1.8 \ (\text{kg · m/min})$$

$$\frac{1565}{1.8} = \frac{1.8}{1.8} \ \text{kg · m/min}$$

$$870 = \text{kg·m/min}$$

$$\text{watts} = \frac{\text{kg · m/min}}{6.12}$$

$$\frac{870}{6.12} = \textbf{142 watts}$$

Problem 4

$$\text{METs} = \frac{\dot{V}O_2 \ mL/kg/min}{3.5}$$

$$= \frac{24}{3.5}$$

$$= 6.9 \ \text{METs}$$

Determine work rate based on $\dot{V}O_2$:

$$\dot{V}O_2 \ (mL/kg/min) = 1.8 \ \frac{\text{work rate (kg · m/min)}}{\text{body mass (kg)}} + 7$$

$$24 = 1.8 \ \frac{\text{kg · m/min}}{95} + 7$$

$$-7 \qquad\qquad\qquad -7$$

$$17 = 1.8 \left(\frac{\text{kg} \cdot \text{m/min}}{95} \right)$$

$$\cdot \, 95 = \cdot \, 95$$

$$1615 = 1.8 \, (\text{kg} \cdot \text{m/min})$$

$$\frac{1615}{1.8} = \frac{1.8}{1.8} \, \text{kg} \cdot \text{m/min}$$

$$897 = \text{kg} \cdot \text{m/min}$$

$$\text{watts} = \frac{\text{kg} \cdot \text{m/min}}{6.12}$$

$$\frac{897}{6.12} = \mathbf{146.6 \ watts}$$

Determine resistance based on kg · m/min:

$$\text{work rate (kg} \cdot \text{m/min)} = R \, (\text{kg}) \times D \, (\text{m}) \times \text{rev/min}$$

$$897 = R \cdot 6 \cdot 50$$

$$897 = R \cdot 300$$

$$\frac{897}{300} = \frac{300}{300}$$

$$\mathbf{2.99 = R}$$

Determine kcals/min (must convert $\dot{V}O_2$ into L/min):

$$24 \, \text{mL/kg/min} \cdot 95 \, \text{kg}$$

$$= 2280 \, \text{mL/min}$$

$$\frac{2280}{1000}$$

$$= 2.28 \, \text{L/min}$$

$$\text{kcals/min} = \dot{V}O_2 \, \text{L/min} \cdot 5$$

$$= 2.28 \cdot 5$$

$$= 11.4 \, \text{kcals/min}$$

Problem 5

Determine work rate in kg · m/min:

$$\text{kg} \cdot \text{m/min} = \text{watts} \cdot 6.12$$

$$= 100 \cdot 6.12$$

$$= 612 \, \text{kg} \cdot \text{m/min}$$

Determine resistance based upon kg · m/min:

$$\text{work rate (kg} \cdot \text{m/min)} = R \text{ (kg)} \cdot D \text{ (m)} \cdot \text{rev/min}$$

$$612 = R \cdot 6 \cdot 65$$

$$612 = R \cdot 390$$

$$\frac{612}{390} = R\,\frac{390}{390}$$

$$1.6 \text{ kg} = R$$

Determine $\dot{V}O_2$:

$$\dot{V}O_2 \text{ (mL/kg/min)} = 1.8\,\frac{\text{work rate (kg} \cdot \text{m/min)}}{\text{body mass (kg)}} + 7$$

$$= 1.8\,\frac{612}{70} + 7$$

$$= 1.8 \cdot 8.74 + 7$$

$$= 15.7 + 7$$

$$= 22.7 \text{ mL/kg/min}$$

$$\text{METs} = \frac{\dot{V}O_2 \text{ mL/kg/min}}{3.5}$$

$$\frac{22.7}{3.5}$$

$$= 6.5 \text{ METs}$$

Determine kcals/min (must convert $\dot{V}O_2$ into L/min):

$$22.7 \text{ mL/kg/min} \cdot 70 \text{ kg}$$

$$= 1589 \text{ mL/min}$$

$$\frac{1589}{1000}$$

$$= 1.59 \text{ L/min}$$

$$\text{kcalsmin} = \dot{V}O_2 \text{ L/min} \cdot 5$$

$$= 1.59 \cdot 5$$

$$= 8.0 \text{ kcals/min}$$

Problem 6

Determine watts:

$$\text{watts} = \frac{\text{kg} \cdot \text{m/min}}{6.12}$$

$$\frac{700}{6.12} = 114 \text{ watts}$$

Determine resistance based upon kg · m/min:

$$\text{work rate (kg · m/min)} = R \text{ (kg)} \cdot D \text{ (m)} \cdot \text{rev/min}$$

$$700 = R \cdot 6 \cdot 60$$

$$700 = R \cdot 360$$

$$\frac{700}{360} = R \frac{360}{360}$$

$$1.94 \text{ kg} = R$$

Determine $\dot{V}O_2$:

$$\dot{V}O_2 \text{ (mL/kg/min)} = 1.8 \frac{\text{work rate (kg · m/min)}}{\text{body mass (kg)}} + 7$$

$$= 1.8 \left(\frac{700}{80} \right) + 7$$

$$= 1.8 \cdot 8.75 + 7$$

$$= 15.75 + 7$$

$$= 22.75 \text{ mL/kg/min}$$

$$\text{METs} = \frac{\dot{V}O_2 \text{ mL/kg/min}}{3.5}$$

$$= \frac{22.75}{3.5}$$

$$= 6.5 \text{ METs}$$

Determine kcals/min (must convert $\dot{V}O_2$ into L/min):

$$22.75 \text{ mL/kg/min} \cdot 80 \text{ kg}$$

$$= 1820 \text{ mL/min}$$

$$\frac{1820}{1000}$$

$$= 1.8 \text{ L/min}$$

$$\text{kcals/min} = \dot{V}O_2 \text{ L/min} \cdot 5$$

$$= 1.8 \cdot 5$$

$$= 9.0 \text{ kcals/min}$$

Practice Table 2

Problem	$\dot{V}O_2$ (mL/kg/min)	Body mass (kg)	Kg · m/min	Watts	Running speed (no grade) miles/hr
1	35.5	70	1089	178	5.9
2	41.5	75	1438	235	7.1
3	29.2	65	802	131	4.8
Problem	$\dot{V}O_2$ (mL/kg/min)	Body mass (kg)	Kg · m/min	Watts	Walking speed (no grade) miles/hr
1	12	70	194	32	3.2
2	10	75	125	20	2.4
3	11.0	65	144	24	2.8

8

ARM ERGOMETER EQUATION

The arm ergometer is a useful device for both exercise testing and training. In cases in which an individual is unable to perform lower body exercise, or when you wish to determine arm-specific aerobic power, the arm ergometer is used. Be aware that during arm exercise, due to the reduced muscle mass involved, blood pressure will be higher at a given heart rate than during lower body exercise. In addition, work rates assigned will be much lower than for leg ergometer exercise.

DERIVATION OF THE ARM ERGOMETER EQUATION

In arm ergometer exercise, the client is pedaling against a resistance that moves a flywheel a given distance with each turn of the pedal crank. Computing the number of pedal turns in a minute helps us compute the work rate. Work rate during arm ergometer exercise is computed in kilogram meters per minute $(kg \cdot m/min)$:

$$\text{work rate } (kg \cdot m \cdot min^{-1}) = R \text{ (kg)} \times D \text{ (m)} \times \text{rev/min}$$

Where R = Resistance or load, expressed as kilograms
       D = Distance the flywheel travels with each pedal crank
       Rev/min = revolutions (pedal cranks) per minute

A Note on D

The flywheel on an arm ergometer may vary by manufacturer; however, the Monark arm ergometer is commonly used. This will be the ergometer on which the following examples will be based.

Monark arm ergometers have a flywheel that moves 2.4 m per revolution.

Be sure to identify the type of cycle ergometer being used, so that you can accurately quantify work rate.

A Note on Work Rate

Another way to express work rate is in watts. These units are commonly used in health and fitness settings, as well as in the research literature. Conversion from kg · m/min to watts is quite simple:

$$\text{watts} = \frac{\text{kg} \cdot \text{m/min}}{6.12} \quad \text{or} \quad \text{kg} \cdot \text{m/min} = \text{watts} \cdot 6.12$$

OXYGEN COST OF ARM ERGOMETER EXERCISE

Arm ergometer exercise involves three components when calculating the oxygen cost ($\dot{V}O_2$) of work.

- The oxygen cost of unloaded cycling is negligible, and therefore not included in the equation.
- The external resistance or load placed on the flywheel is approximately 3.0 mL/kg/min for each kg · m/min. This is higher than for the leg ergometer, and is due to the increased use of trunk stabilization muscles used during the cranking as well as reduced efficiency or arm-versus leg-cycling.
- Resting oxygen consumption, which is approximately 3.5 mL/kg/min.

THE ACSM ARM ERGOMETER EQUATION

The ACSM arm ergometer equation follows. This equation is most accurate for work rates between 150 and 750 kgm · min^{-1} (25–125 watts).

$$\dot{V}O_2 \ (\text{mL/kg/min}) = 3.0 \ \frac{\text{kg} \cdot \text{m/min}}{\text{body mass (kg)}} + 3.5$$

SPECIFIC USES OF THE ARM ERGOMETER EQUATION

Determine the Caloric Cost of Arm Ergometer Exercise

For an individual who is cycling, you can compute first the oxygen cost ($\dot{V}O_2$) and then determine the caloric cost.

Example: An individual is cycling at a Resistance (R) of 1.0 kg on a Monark arm ergometer. He is pedaling 50 rev/min and weighs 90 kg.

Step 1: Determine work rate in kg · m/min:

$$\text{work rate (kg} \cdot \text{m/min)} = R \ (\text{kg}) \cdot D \ (\text{m}) \cdot \text{rev/min}$$
$$R= 1.0 \text{ kg } D = 2.4 \text{ m (because it is a Monark arm erg) rev/min} = 50$$
$$\text{work rate (kg} \cdot \text{m/min)} = 1.0 \cdot 2.4 \cdot 50$$
$$= \textbf{120 kg} \cdot \textbf{m/min}$$

Step 2: Calculate $\dot{V}O_2$ (remember that body mass (BM) is 90 kg):

$$\dot{V}O_2 \text{ (mL/kg/min)} = 3.0 \frac{\text{kg} \cdot \text{m/min}}{\text{body mass (kg)}} + 3.5$$

$$= 3.0 \left(\frac{120}{90} \right) + 3.5$$

$$= 3.0 \, (1.3) + 3.5$$

$$= 3.9 + 3.5$$

$$= \mathbf{7.4 \ mL/kg/min}$$

Step 3: Convert $\dot{V}O_2$ in mL/kg/min into kcals/min:

$$7.4 \text{ mL/kg/min} \cdot 90 \text{ kg} = 666 \text{ mL/min}$$

$$\frac{666}{1000} = 0.67 \text{ L/min}$$

$$0.67 \cdot 5 = \mathbf{3.35 \ kcals/min}$$

Determine the Appropriate Work Rate for an Exercise Prescription

In many cases, you have a training $\dot{V}O_2$, and need to compute work rate in kg · m/min or perhaps even watts.

Example: A 50 kg woman has a training $\dot{V}O_2$ of 10 ml/kg/min. Compute the work rate at which she should train in both watts and kg · m/min.

Step 1: Determine kg · m/min:

$$\dot{V}O_2 \text{ (mL/kg/min)} = 3.0 \frac{\text{kg} \cdot \text{m/min}}{\text{body mass (kg)}} + 3.5$$

$$10.0 = 3.0 \frac{\text{kg} \cdot \text{m/min}}{50} + 3.5$$

$$-3.5 \qquad\qquad\qquad -3.5$$

$$6.5 = 3.0 \frac{\text{kg} \cdot \text{m/min}}{50}$$

$$\cdot\, 50 \qquad\qquad\qquad \cdot 50$$

$$325 = 3.0 \, (\text{kg} \cdot \text{m/min})$$

$$\frac{325}{3.0} = \frac{3.0}{3.0} \, (\text{kg} \cdot \text{m/min})$$

$$108 = \mathbf{kg \cdot m/min}$$

Step 2: Convert to watts:

$$\text{watts} = \frac{\text{kg} \cdot \text{m/min}}{6.12}$$

$$\text{watts} = \frac{108}{6.12}$$

watts = 17.6

PRACTICE TABLE 1. *Fill in the appropriate information and complete the table (answers on page 63). Fill in the blanks using the following equations:*

$$\dot{V}O_2 \text{ (mL/kg/min)} = 3.0 \frac{\text{kg} \cdot \text{m/min}}{\text{body mass (kg)}} + 3.5$$

$$\text{work rate (kg} \cdot \text{m/min)} = R \text{ (kg)} \cdot D \text{ (m)} \cdot \text{rev/min}$$

$$\text{watts} = \frac{\text{kg} \cdot \text{m/min}}{6.12} \quad \text{or} \quad \text{kg} \cdot \text{m/min} = \text{watts} \cdot 6.12$$

$$\text{kcals/min} = \dot{V}O_2 \text{ L/min} \cdot 5 \qquad \text{METs} = \frac{\dot{V}O_2 \text{ mL/kg/min}}{3.5}$$

Problem	V̇O₂ ml/ kg/min	METS	Body Wt. (kg)	kcals/ min	R (kg)	Rev/ min	Watts	Kg · m/ min
1			59		1.5	55		
2			70		2.0	60		
3			85	6		50		
4		5	54			65		
5			95	3		60		
6	12.0		47			50		

Note: Assumes a Monark arm ergometer, 2.4 m/rev

SOLUTIONS

Problem 1

Determine work rate:

$$\text{work rate (kg} \cdot \text{m/min)} = R \text{ (kg)} \cdot D \text{ (m)} \cdot \text{rev/min}$$

$$= 1.5 \cdot 2.4 \cdot 55$$

$$= \textbf{198 kg} \cdot \textbf{m/min}$$

$$\text{watts} = \frac{\text{kg} \cdot \text{m/min}}{6.12}$$

$$= \frac{198}{6.12}$$

$$= \textbf{32 watts}$$

Determine $\dot{V}O_2$ using arm ergometer equation:

$$\dot{V}O_2 \ (\text{mL/kg/min}) = 3.0 \ \frac{\text{kg} \cdot \text{m/min}}{\text{body mass (kg)}} + 3.5$$

$$= 3.0 \left(\frac{198}{59}\right) + 3.5$$

$$= 3.0 \ (3.36) + 3.5$$

$$= 10.08 + 3.5$$

$$= \mathbf{13.58 \ mL/kg/min}$$

$$\text{METs} = \frac{\dot{V}O_2 \, \text{mL/kg/min}}{3.5}$$

$$= \frac{13.58}{3.5}$$

$$= \mathbf{3.9 \ METs}$$

Determine kcals expended (must convert $\dot{V}O_2$ into L/min):

$$13.58 \ \text{mL/kg/min} \cdot 59 \ \text{kg} = 801 \ \text{mL/min}$$

$$\frac{801}{1000} = 0.801 \ \text{L/min}$$

$$\text{kcals/min} = \dot{V}O_2 \ \text{L} \cdot \text{min} \cdot 5$$

$$= 0.801 \cdot 5$$

$$= \mathbf{4.0 \ kcals/min}$$

Problem 2

Determine work rate:

$$\text{work rate (kg} \cdot \text{m/min)} = R \ (\text{kg}) \cdot D \ (\text{m}) \cdot \text{rev/min}$$

$$= 2.0 \cdot 2.4 \cdot 60$$

$$= \mathbf{288 \ kg \cdot m/min}$$

$$\text{watts} = \frac{288}{6.12}$$

$$= \mathbf{47 \ watts}$$

Determine $\dot{V}O_2$ using arm ergometer equation:

$$\dot{V}O_2 \ (\text{mL/kg/min}) = 3.0 \ \frac{\text{kg} \cdot \text{m/min}}{\text{body mass (kg)}} + 3.5$$

$$= 3.0 \left(\frac{288}{70}\right) + 3.5$$

$$= 3.0 \ (4.11) + 3.5$$

$$= 12.33 + 3.5$$

$$= \mathbf{15.83 \ mL/kg/min}$$

PRACTICE TABLE 1.

Problem	$\dot{V}O_2$ ml/ kg/min	METs	Body wt. (kg)	kcals/ min	R (kg)	Rev/ min	Watts	Kg · m/ min
1	13.6	3.9	59	4.0	1.5	55	32	198
2	15.8	4.5	70	5.6	2.0	60	47	288
3	14.1	4.0	85	6.0	2.5	50	49	300
4	17.5	5.0	54	4.7	1.6	65	41	252
5	6.3	1.8	95	3.0	0.6	60	15	89
6	12.0	3.4	47	2.8	1.1	50	22	133

Note: Assumes a Monark arm ergometer, 2.4 m/rev

$$\text{METs} = \frac{\dot{V}O_2\,(\text{mL/kg/min})}{3.5}$$

$$= \frac{15.83}{3.5}$$

$$= 4.5\ \text{METs}$$

Determine kcals expended (must convert $\dot{V}O_2$ into L/min):

$$15.83\ \text{mL/kg/min} \cdot 70\ \text{kg} = 1108.1\ \text{mL/min}$$

$$\frac{1108.1}{1000} = 1.11\ \text{L/min}$$

$$\text{kcals} = \dot{V}O_2\ \text{L/min} \cdot 5$$

$$= 1.11 \cdot 5$$

$$= 5.6\ \text{kcals/min}$$

Problem 3

Convert kcals/min into $\dot{V}O_2$:

$$\frac{6\ \text{kcals}}{5} = 1.2\ \text{L/min}$$

$$1.2 \cdot 1000 = 1200\ \text{mL/min}$$

$$\frac{1200}{85\ \text{kg}} = 14.1\ \text{mL/kg/min}$$

$$\text{METs} = \frac{\dot{V}O_2\,(\text{mL/kg/min})}{3.5}$$

$$= \frac{14.1}{3.5}$$

$$= 4.0\ \text{METs}$$

Convert $\dot{V}O_2$ into work rate using arm ergometer equation:

$$\dot{V}O_2 \text{ (mL/kg/min)} = 3.0 \frac{\text{kg} \cdot \text{m/min}}{\text{body mass (kg)}} + 3.5$$

$$14.1 = 3.0 \frac{\text{kg} \cdot \text{m/min}}{85} + 3.5$$

$$-3.5 \qquad\qquad\qquad -3.5$$

$$10.6 = 3.0 \frac{\text{kg} \cdot \text{m/min}}{85}$$

$$\cdot 85 \qquad\qquad\qquad \cdot 85$$

$$901 = 3.0 \,(\text{kg} \cdot \text{m/min})$$

$$\frac{901}{3.0} = \frac{3.0}{3.0} \,(\text{kg} \cdot \text{m/min})$$

$$300 = \textbf{kg} \cdot \textbf{m/min}$$

$$\text{watts} = \frac{\text{kg} \cdot \text{m/min}}{6.12}$$

$$= \frac{300}{6.12}$$

$$= \textbf{49 watts}$$

Determine load based on work rate:

$$\text{work rate (kg} \cdot \text{m/min)} = \text{R (kg)} \cdot \text{D (m)} \cdot \text{rev/min}$$

$$300 = \text{R} \cdot 2.4 \cdot 50$$

$$300 = \text{R} \cdot 120$$

$$\frac{300}{120} = \text{R} \frac{120}{120}$$

$$\textbf{2.5 kg = R}$$

Problem 4

Determine $\dot{V}O_2$:

$$5 \text{ METs} \cdot 3.5 = 17.5 \text{ mL/kg/min}$$

Determine kcals expended (must convert $\dot{V}O_2$ into L/min):

$$17.5 \text{ mL/kg/min} \cdot 54\text{kg} = 945 \text{ mL/min}$$

$$\frac{945}{1000} = 0.945 \text{ L/min}$$

$$\text{kcals/min} = \dot{V}O_2 \text{ L/min} \cdot 5$$

$$= 0.945 \cdot 5$$

$$= \textbf{4.7 kcals/min}$$

Convert $\dot{V}O_2$ into work rate using arm ergometer equation:

$$\dot{V}O_2 \text{ (mL/kg/min)} = 3.0 \frac{\text{kg} \cdot \text{m/min}}{\text{body mass (kg)}} + 3.5$$

$$17.5 = 3.0 \frac{\text{kg} \cdot \text{m/min}}{54} + 3.5$$

$$-3.5 \qquad\qquad\qquad -3.5$$

$$14 = 3.0 \frac{\text{kg} \cdot \text{m/min}}{54}$$

$$\cdot 54 \qquad\qquad\qquad \cdot 54$$

$$756 = 3.0 \text{ kg} \cdot \text{m/min}$$

$$\frac{756}{3.0} = \frac{3.0}{3.0} \text{ kg} \cdot \text{m/min}$$

$$\mathbf{252 = kg \cdot m/min}$$

$$\text{watts} = \frac{\text{kg} \cdot \text{m/min}}{6.12}$$

$$= \frac{252}{6.12}$$

$$= \mathbf{41 \ watts}$$

Determine load based on work rate:

$$\text{work rate (kg} \cdot \text{m/min)} = R \text{ (kg)} \cdot D \text{ (m)} \cdot \text{rev/min}$$

$$252 = R \cdot 2.4 \cdot 65$$

$$252 = R \cdot 156$$

$$\frac{252}{156} = R \frac{156}{156}$$

$$\mathbf{1.6 \ kg = R}$$

Problem 5

Convert kcals/min into $\dot{V}O_2$:

$$\frac{3 \text{ kcals}}{5} = 0.6 \text{ L/min}$$

$$0.6 \cdot 1000 = 600 \text{ mL/min}$$

$$\frac{600}{95} = \mathbf{6.3 \ ml/kg/min}$$

$$\text{METs} = \frac{\dot{V}O_2 \ (\text{mL/kg/min})}{3.5}$$

$$= \frac{6.3}{3.5}$$

$$= \textbf{1.8 METs}$$

Convert $\dot{V}O_2$ into work rate using arm ergometer equation:

$$\dot{V}O_2 \ \text{ml/kg/min} = 3.0 \ \frac{\text{kg} \cdot \text{m/min}}{\text{body mass (kg)}} + 3.5$$

$$6.3 = 3.0 \ \frac{\text{kg} \cdot \text{m/min}}{95} + 3.5$$

$$-3.5 \qquad\qquad\qquad -3.5$$

$$2.8 = 3.0 \ \frac{\text{kg} \cdot \text{m/min}}{95}$$

$$\cdot \ 95 \qquad\qquad\qquad \cdot \ 95$$

$$266 = 3.0 \ (\text{kg} \cdot \text{m/min})$$

$$\frac{266}{3.0} = \frac{3.0}{3.0} \ \text{kg} \cdot \text{m/min}$$

$$\textbf{89} = \textbf{kg} \cdot \textbf{m/min}$$

$$\text{watts} = \frac{\text{kg} \cdot \text{m/min}}{6.12}$$

$$= \frac{89}{6.12}$$

$$= \textbf{14.5 watts}$$

Determine load based on work rate:

$$\text{work rate (kg} \cdot \text{m/min)} = \text{R (kg)} \cdot \text{D (m)} \cdot \text{rev/min}$$

$$89 = \text{R} \cdot 2.4 \cdot 60$$

$$89 = \text{R} \cdot 144$$

$$\frac{89}{144} = \text{R} \ \frac{144}{144}$$

$$\textbf{0.62 kg} = \textbf{R}$$

Problem 6

$$\text{METs} = \frac{\dot{V}O_2 \ \text{mL/kg/min}}{3.5}$$

$$= \frac{12.0}{3.5}$$

$$= \textbf{3.4 METs}$$

Determine kcals/min expended (must convert $\dot{V}O_2$ into L/min):

$$12.0 \text{ mL/kg/min} \cdot 47 \text{ kg} = 564 \text{ mL/min}$$

$$\frac{564}{1000} = 0.564 \text{ L/min}$$

$$
\begin{aligned}
\text{kcals/min} = \dot{V}O_2 \text{ L/min} \cdot 5 \\
= 0.564 \cdot 5 \\
= \textbf{2.8 kcals/min}
\end{aligned}
$$

Convert $\dot{V}O_2$ into work rate using arm ergometer equation:

$$\dot{V}O_2 \text{ (mL/kg/min)} = 3.0 \frac{\text{kg} \cdot \text{m/min}}{\text{body mass (kg)}} + 3.5$$

$$12.0 = 3.0 \frac{\text{kg} \cdot \text{m/min}}{47} + 3.5$$

$$-3.5 \qquad\qquad\qquad -3.5$$

$$8.5 = 3.0 \frac{\text{kg} \cdot \text{m/min}}{47}$$

$$\cdot 47 \qquad\qquad\qquad \cdot 47$$

$$399.5 = 3.0 \text{ (kg} \cdot \text{m/min)}$$

$$\frac{399.5}{3.0} = \frac{3.0}{3.0} \text{ (kg} \cdot \text{m/min)}$$

$$\textbf{133} = \textbf{kg} \cdot \textbf{m/min}$$

$$\text{watts} = \frac{\text{kg} \cdot \text{m/min}}{6.12}$$

$$= \frac{133}{6.12}$$

$$= \textbf{21.7 watts}$$

Determine load based on work rate:

$$\text{work rate (kg} \cdot \text{m/min)} = R \text{ (kg)} \cdot D(m) \cdot \text{rev/min}$$

$$133 = R \cdot 2.4 \cdot 50$$

$$133 = R \cdot 120$$

$$\frac{133}{120} = R \frac{120}{120}$$

$$\textbf{1.1 kg} = \textbf{R}$$

9

STEPPING EQUATION

Stepping is used as both a mode of exercise testing as well as a form of rhythmic exercise. Step height is varied, but stepping typically involves a four-step process to complete a stepping cycle. One step is stepping forward and onto a box or bench, lifting the body and the other leg onto the box, then stepping backward and down, followed by the procedure with the second leg. Oxygen cost is dependent upon step rate and step height.

DERIVATION OF THE STEPPING EQUATION

Similar to that for walking up an incline, the stepping equation is composed of a horizontal component, a vertical component, and a resting component:

1. The horizontal component represents the cost of stepping forward and backward onto the box or bench. The oxygen cost of the complete cycle of stepping forward and back during the stepping cycle is approximately 0.2 mL per kilogram body mass (0.2 mL/kg/min).
2. The vertical component represents the oxygen cost of both stepping up (1.8 mL/kg/min) and stepping down (1.33 mL/kg/min).
3. The resting component is the same as seen in the walking and running equations: 3.5 mL/kg/min.

The accuracy of the stepping equation is highest between stepping rates of 12 and 30 steps per minute and step heights of 0.04 and 0.40 meters (1.6 to 15.7 inches).

THE ACSM STEPPING EQUATION

The ACSM stepping equation is

$$\dot{V}O_2 \text{ (mL/kg/min)} = (0.2 \cdot f) + (1.33 \cdot 1.88 \cdot H \cdot f) + 3.5$$
$$f = \text{stepping rate}$$
$$H = \text{step height in meters}$$

SPECIFIC USES OF THE STEPPING EQUATION

Determine The Caloric Cost of Stepping Exercise

When individuals use stepping as a form of exercise (i.e., step aerobics) you can calculate the energy cost of the work if you know the stepping rate and height of the step.

Example: An individual stepping to a beat of 25 steps per minute on a step that is 7 inches high. What is the oxygen cost?

Step 1: Convert step height from inches to meters.

One inch equals 2.54 cm: 7 in · 2.54 = 17.78 cm
One meter equals 100 cm: 17.78 cm/100 = 0.178 m

Step 2: Determine $\dot{V}O_2$ using the stepping equation.

$$\dot{V}O_2 \ (mL/kg/min) = (0.2 \cdot f) + (1.33 \cdot 1.88 \cdot H \cdot f) + 3.5$$
$$= (0.2 \cdot 25) + (1.33 \cdot 1.88 \cdot 0.178 \cdot 25) + 3.5$$
$$= 5 + 11.1 + 3.5$$
$$= \mathbf{19.6 \ mL/kg/min}$$

Determine a Stepping Rate Based Upon a $\dot{V}O_2$ Prescription

When the training $\dot{V}O_2$ is known, you can use an available step (with a known step height) and prescribe a step rate using the stepping equation:

Example: A client is prescribed an exercise intensity of 7 METs. Using a 10-inch step, calculate the desired step rate needed to equate to the prescribed MET intensity.

Step 1: Convert METs into $\dot{V}O_2$ mL/kg/min.

$$7 \ METs \cdot 3.5 = 24.5 \ mL/kg/min$$

Step 2: Convert step height from inches to meters.

One inch equals 2.54 cm: 7 in · 2.54 = 25.4 cm
One meter equals 100 cm: 25.4 cm/100 = 0.254 m

Step 3: Use the stepping equation and solve for step rate.

$$\dot{V}O_2 \ (mL/kg/min) = (0.2 \cdot f) + (1.33 \cdot 1.88 \cdot H \cdot f) + 3.5$$
$$24.5 = (0.2 \cdot f) + (1.33 \cdot 1.88 \cdot 0.254 \cdot f) + 3.5$$
$$-3.5 \hspace{4cm} -3.5$$
$$21 = (0.2 \cdot f) + (0.635 \cdot f)$$
$$21 = 0.835 \cdot f)$$

$$\frac{21}{0.835} = \frac{0.835}{0.835} \cdot f$$
$$25.1 = f$$

Answer: A step rate of 25 steps per minute on a 10-inch step will elicit a 7 MET intensity.

PRACTICE TABLE

PRACTICE TABLE 1. *Fill in the appropriate information and complete the table (answers start below). Fill in the blanks using the following equations:*

$$\dot{V}O_2 \text{ (mL/kg/min)} = (0.2 \cdot f) + (1.33 \cdot 1.88 \cdot H \cdot f) + 3.5$$

One inch equals 2.54 cm   One meter equals 100 cm

$$\text{kcals/min} = \dot{V}O_2 \text{ L/min} \cdot 5 \qquad \text{METs} = \frac{\dot{V}O_2 \text{ (mL/kg/min)}}{3.5}$$

Problem	$\dot{V}O_2$ (mL/kg/min)	METs	Body mass (kg)	kcals/min	Step rate (steps/min)	Step height (inches)
1			54		20	8
2			67		25	15
3		8	72			10
4			94	10	30	
5	21.0		66			12
6			85		12	9

SOLUTIONS

Problem 1

Convert step height to meters:

One inch equals 2.54 cm   One meter equals 100 cm
8 in · 2.54 = 20.32 cm   20.32 cm/100 = 0.20 m

Solve for $\dot{V}O_2$ using the stepping equation:

$$\dot{V}O_2 \text{ (mL/kg/min)} = (0.2 \cdot f) + (1.33 \cdot 1.88 \cdot H \cdot f) + 3.5$$
$$= 0.2 \, (20) + (1.33 \cdot 1.88 \cdot 0.20 \cdot 20) + 3.5$$
$$= 4 + 10 + 3.5$$
$$= \mathbf{17.05 \ mL/kg/min}$$

ANSWERS TO PRACTICE TABLE

PRACTICE TABLE 1.

Problem	$\dot{V}O_2$ (mL/kg/min)	METs	Body mass (kg)	kcals/min	Step rate (steps/min)	Step height (inches)
1	17.25	5	54	4.7	20	8
2	32.3	9.2	67	11	25	15
3	28.0	8.0	72	10.0	30	10
4	21.3	6.1	94	10.0	30	6.7
5	21.0	6.0	66	7.0	18	12
6	12.8	3.7	85	5.5	12	9

Convert to METs:

$$\frac{17.05}{3.5} = \textbf{5.0 METs}$$

Determine kcals/min (must convert $\dot{V}O_2$ into L/min):

$$17.06 \text{ mL/kg/min} \cdot 54 \text{ kg} = 945 \text{ mL/min}$$

$$\frac{945}{1000} = 0.945 \text{ L/min}$$

$$\text{kcals/min} = \dot{V}O_2 \text{ L/min} \cdot 5$$
$$= 0.945 \cdot 5$$
$$= \textbf{4.7 kcals/min}$$

Problem 2

Convert step height to meters:

One inch equals 2.54 cm   One meter equals 100 cm
15 in · 2.54 = 38.1 cm   38.1/100 = 0.38 m

Solve for $\dot{V}O_2$ using the stepping equation:

$$\dot{V}O_2 \text{ (mL/kg/min)} = (0.2 \cdot f) + (1.33 \cdot 1.88 \cdot H \cdot f) + 3.5$$
$$= 0.2\,(25) + (1.33 \cdot 1.88 \cdot 0.38 \cdot 25) + 3.5$$
$$= 5 + 23.8 + 3.5$$
$$= \textbf{32.3 mL/kg/min}$$

Convert to METs:

$$\frac{32.3}{3.5} = \textbf{9.2 METs}$$

Determine kcals/min (must convert $\dot{V}O_2$ into L/min):

$$32.3 \ (mL/kg/min \cdot 67 \ kg) = 2164 \ mL/min$$

$$\frac{2164 \ mL/min}{1000} = 2.2 \ L/min$$

$$kcals/min = \dot{V}O_2 \ (L/min \cdot 5)$$
$$= 2.2 \cdot 5$$
$$= \mathbf{11 \ kcals/min}$$

Problem 3

Convert step height to meters:

One inch equals 2.54 cm   One meter equals 100 cm

10 in $\cdot$ 2.54 = 25.4 cm   25.4/100 = 0.25 m

Convert METs to $\dot{V}O_2$:

$$8 \ METs \cdot 3.5 = 28.0 \ mL/kg/min$$

Determine kcals/min (must convert $\dot{V}O_2$ into L/min):

$$28.0 \ mL/kg/min \cdot 72 \ kg = 2016 \ mL/min$$

$$\frac{2016 \ mL/min}{1000} = 2.0 \ L/min$$

$$kcals/min = \dot{V}O_2 \ L/min \cdot 5$$
$$= 2.0 \cdot 5$$
$$= \mathbf{10.0 \ kcals/min}$$

Use the stepping equation to solve for step rate (f):

$$\dot{V}O_2 \ (mL/kg/min) = (0.2 \cdot f) + (1.33 \cdot 1.88 \cdot H \cdot f) + 3.5$$
$$28.0 = (0.2 \cdot f) + (1.33 \cdot 1.88 \cdot 0.25 \cdot f) + 3.5$$
$$-3.5 \qquad\qquad\qquad\qquad\qquad\qquad -3.5$$
$$24.5 = 0.2 \ f + 0.63 \ f$$
$$24.5 = 0.83 \ f$$
$$\frac{24.5}{0.83} = \frac{0.83}{0.83} f$$
$$\mathbf{29.5 \ steps/min = f}$$

Problem 4

Convert kcals/min to $\dot{V}O_2$:

$$\frac{10 \ kcals/min}{5} = 2.0 \ L/min$$

$$2.0 \cdot 1000 = 2000 \ mL/min$$

$$\frac{2000 \ mL/min}{94} = \mathbf{21.3 \ mL/kg/min}$$

Convert to METs:

$$\frac{21.3 \text{ mL/kg/min}}{3.5} = 6.1 \text{ METs}$$

Use the stepping equation to solve for step height (H):

$$\dot{V}O_2 \text{ (mL/kg/min)} = (0.2 \cdot f) + (1.33 \cdot 1.88 \cdot H \cdot f) + 3.5$$
$$21.3 = (0.2 \cdot 30) + (1.33 \cdot 1.88 \cdot H \cdot 30) + 3.5$$
$$21.3 = 6 + 75.0 \text{ H} + 3.5$$
$$21.3 = 75.0 \text{ H} + 9.5$$
$$-9.5 \qquad\qquad -9.5$$
$$11.8 = 75.0 \text{ H}$$
$$\frac{11.8}{75.0} = \frac{75.0}{75.0} \text{ H}$$
$$\mathbf{0.17 \text{ meter} = H}$$

Convert step height into inches:

$$0.17 \text{ m} \cdot 100 = 17 \text{ cm}$$
$$\frac{17 \text{ cm}}{2.54} = 6.7 \text{ inches}$$

Problem 5

Convert step height to meters:

One inch equals 2.54 cm   One meter equals 100 cm
12 in · 2.54 = 30.5 cm   30.5 cm/100 = 0.31 m

Determine kcals/min (must convert $\dot{V}O_2$ into L/min):

$$21.0 \text{ mL/kg/min} \cdot 66 \text{ kg} = 1386 \text{ mL/min}$$
$$\frac{1386 \text{ mL/min}}{1000} = 1.4 \text{ L/min}$$

$$\text{kcals/min} = \dot{V}O_2 \text{ L/min} \cdot 5$$
$$= 1.4 \cdot 5$$
$$= \mathbf{7.0 \text{ kcalsmin}}$$

Determine METs:

$$\frac{21.0 \text{ mL/kg/min}}{3.5} = 6 \text{ METs}$$

Use the stepping equation to solve for step rate (f):

$$\dot{V}O_2 \text{ (mL/kg/min)} = (0.2 \cdot f) + (1.33 \cdot 1.88 \cdot H \cdot f) + 3.5$$
$$21.0 = 0.2(\text{SR}) + (1.33 \cdot 1.88 \cdot 0.31 \cdot f) + 3.5$$
$$-3.5 \qquad\qquad\qquad -3.5$$
$$17.5 = 0.2 \text{ f} + 0.78 \text{ f}$$
$$17.5 = 0.98 \text{ f}$$

$$\frac{17.5}{0.98} = \frac{0.98}{0.98}\,f$$

17.9 steps/min = f

Problem 6

Convert step height to meters:

One inch equals 2.54 cm   One meter equals 100 cm
9 in · 2.54 = 22.9 cm   22.9 cm/100 = .23 m

Solve for $\dot{V}O_2$ using the stepping equation:

$$\dot{V}O_2 \text{ (mL/kg/min)} = (0.2 \cdot f) + (1.33 \cdot 1.88 \cdot H \cdot f) + 3.5$$
$$= 0.2\,(12) + (1.33 \cdot 1.88 \cdot 0.23 \cdot 12) + 3.5$$
$$= 2.4 + 6.9 + 3.5$$
$$= \mathbf{12.8\ mL/kg/min}$$

Convert to METs:

$$\frac{12.8\text{ mL/kg/min}}{3.5} = \textbf{3.7 METs:}$$

Determine kcals/min (must convert $\dot{V}O_2$ into L/min):

$$12.8 \text{ mL/kg/min} \cdot 85 \text{ kg} = 1088 \text{ mL/min}$$
$$\frac{1088 \text{ mL/min}}{1000} = 1.09 \text{ L/min}$$

$$\text{kcals/min} = \dot{V}O_2 \text{ L/min} \cdot 5$$
$$= 1.09 \cdot 5$$
$$= \mathbf{5.5\ kcals/min}$$

PRACTICE PROBLEMS

The following 34 questions will require you to use all of the metabolic equations, as well as conversions to kcals, METs, and other common transformations. The problems are grouped as mini-case studies, with three problems per case. You may need the results from one problem in order to accurately complete the next problem. All problems are worked to their correct solution at the end of the chapter.

A 75 kg man cycles at 120 watts on the leg ergometer.

1. His training $\dot{V}O_2$ is:
   a) 24 METs
   b) 17.06 mL/kg/min
   c) 12.5 mL/kg/min
   d) 7 METs

2. His energy expenditure during a 30-minute ride is:
   a) 277 kcals
   b) 9.2 kcals
   c) 1050 kcals
   d) 300 kcals

3. If he chooses to walk up a 10% grade, what is the appropriate walking speed?
   a) 211 m/min
   b) 3.4 miles/hr
   c) 2.8 miles/hr
   d) 100 m/min

During an aerobic dance class, Lisa, who weighs 58 kg, is stepping on a 9-inch step at a rate of 20 steps per minute.

4. How many kilocalories will she expend in 20 minutes?
   a) 107 kcals
   b) 160 kcals

    c) 5.4 kcals
    d) 185 kcals

5. What is an equivalent work rate on the leg ergometer?
    a) 300 watts
    b) 2264 watts
    c) 120 kg · m/min
    d) 370 kg · m/min

Running on a level track at 9.0 mph, an 80 kg male:

6. Has a $\dot{V}O_2$ of:
    a) 51.7 mL/kg/min
    b) 10 METs
    c) 27.6 mL/kg/min
    d) 42.3 mL/kg/min

7. How many calories per minute is he expending?
    a) 0.11 kcals/min
    b) 50 kcals/min
    c) 20.5 kcals/min
    d) 7.5 kcals/min

John's $\dot{V}O_{2max}$ is 32.0 mL/kg/min. He weighs 74 kg.

8. What is his work rate on the Monark arm ergometer at 40% of $\dot{V}O_{2max}$?
    a) 703 kg · m/min
    b) 688 kg · m/min
    c) 100 kg · m/min
    d) 229 kg · m/min

9. A leg ergometer work rate at 80% of $\dot{V}O_{2reserve}$ is:
    a) 764 kg · m/min
    b) 793 kg · m/min
    c) 195 watts
    d) 110 watts

10. How many calories are expended at an exercise intensity of 60% of $\dot{V}O_{2max}$?
    a) 7.1 kcals/min
    b) 11.8 kcals/min
    c) 7.6 kcals/min
    d) 9.2 kcals/min

Nickole, a 54 kg woman, wants to begin a cross training program.

11.   What is her $\dot{V}O_2$ at 2.5 miles/hr and 12%?
      a) 1525 L/min
      b) 6 METs
      c) 12 METs
      d) 24.7 mL/kg/min

12.   What is the equivalent work rate on the cycle ergometer?
      a) 30 watts
      b) 531 kg · m/min
      c) 575 kg · m/min
      d) 87 kg · m/min

13.   If she steps on a 12-inch step, what step rate is required to elicit the same $\dot{V}O_2$?
      a) 22 steps/min
      b) 20 steps/min
      c) 27 steps/min
      d) 30 steps/min

Cycling on the Monark leg ergometer (flywheel 6 m/rev), Dawn is cycling at 65 rpm against a load of 2.0 kg. She weighs 55 kg.

14.   What is her $\dot{V}O_2$?
      a) 27 mL/kg/min
      b) 36 METs
      c) 1.8 L/min
      d) 920 mL/min

15.   With an estimated $\dot{V}O_{2max}$ of 3.0 L/min you determine a training intensity of 70% of $\dot{V}O_{2reserve}$ for Dawn. This corresponds to a work rate of:
      a) 2.1 kg
      b) 161 watts
      c) 143 watts
      d) 1452 kg · m/min

16.   Dawn's preferred intensity jogging is 6.5 miles/hr. This corresponds to a $\dot{V}O_2$ of:
      a) 34.6 mL/kg/min
      b) 11 METs
      c) 6.5 METs
      d) 2.5 L/min

Steve is working out on the Monark arm ergometer (flywheel 2.4 m/rev). The resistance is set at 1.5 kg, and he is cranking at 60 rpm. He weighs 85 kg.

17. His estimated energy expenditure is:
    a) 4.7 kcals/min
    b) 9.6 kcals/min
    c) 7.4 kcals/min
    d) 8.0 kcals/min

18. Moving to stepping, Steve steps on a 10-inch box at 24 steps /min. His estimated energy expenditure is:
    a) 9.9 kcals/min
    b) 4.7 kcals/min
    c) 11.3 kcals/min
    d) 6.1 kcals/min

19. Completing his circuit, Steve walks on the treadmill at 3.5 miles/hr up a 12% incline. His estimated energy expenditure is:
    a) 11.3 kcals/min
    b) 13.2 kcals/min
    c) 14.1 kcals/min
    d) 9.6 kcals/min

Otto, a client in your cardiac rehab program needs a circuit exercise intensity prescription. His functional capacity is 8 METs. He weighs 90 kg.

20. What is his arm ergometer work rate at 40% of his METs capacity?
    a) 11.2 watts
    b) 73.5 kg · m/min
    c) 75 kg · m/min
    d) 231 kg · m/min

21. When cycling (leg) at 65% of $\dot{V}O_{2reserve}$, how many kilocalories is he expending?
    a) 8.7 kcals/min
    b) 5.5 kcals/min
    c) 19.4 kcals/min
    d) 14.2 kcals/min

22. Walking at 50% of $\dot{V}O_{2max}$, his speed is 3.0 miles/hr. What is the required grade?
    a) 2.5%
    b) 1.7%
    c) 3.0%
    d) 2.0%

Your client, Jim is interested in weight control. He weighs 75 kg.

23. If Jim walks 3.3 miles/hr, how long must he walk to expend the 300 kcals?
     a) 52 min
     b) 42 min
     c) 65 min
     d) 99 min

24. If Jim exercises at an intensity equivalent to 6.0 kcals/min, what is the leg ergometer work rate?
     a) 47 watts
     b) 90 watts
     c) 61 watts
     d) 71 watts

25. Cycling on the arm ergometer at 1.0 kg and 50 rev/min, how long would it take Jim to expend 300 kcals (assume 2.4 m per pedal rev)?
     a) 1 hr 37 min
     b) 52 min
     c) 1 hr
     d) 1 hr 14 min

An exercise test was performed on a triathlete. His $\dot{V}O_{2max}$ was 73.0 mL/kg/min. He weighed 73 kg.

26. His anaerobic threshold was 80% of $\dot{V}O_{2max}$. Determine the running speed at his threshold.
     a) 10 min/mile pace
     b) 58.4 m/sec
     c) 10.8 miles/hr
     d) 200 m/min

27. His leg ergometer work rate is typically 1100 kg · m/min. What percent of $\dot{V}O_{2max}$ is this?
     a) 39% of $\dot{V}O_{2max}$
     b) 47% of $\dot{V}O_{2max}$
     c) 34% of $\dot{V}O_{2max}$
     d) 60% of $\dot{V}O_{2max}$

28. To train for hills, you have him run at a $\dot{V}O_2$ of 50 mL/kg/min up a 5% incline. What is the appropriate running speed?
     a) 190 m/min
     b) 9.0 min/mile pace
     c) 90 m/sec
     d) 9.0 miles/hr

You allow your client, Julie, to self-select training intensity. She weighs 55 kg.

29. If she cycles on the Monark leg ergometer against a 1.5 kg resistance at 75 rev/min, what is her $\dot{V}O_2$?
    a) 22.1 mL/kg/min
    b) 12 METs
    c) 9.8 mL/kg/min
    d) 8.3 METs

30. She wants to replicate the same intensity as the previous question, but by walking on the treadmill up a 10% incline. What is the appropriate walking speed?
    a) 2.5 miles/hr
    b) 4.0 miles/hr
    c) 1.7 miles/hr
    d) 3.4 miles/hr

31. How many kilocalories will she expend during a 30-minute workout?
    a) 240 kcals
    b) 80 kcals
    c) 8.0 kcals
    d) 15.0 kcals

Your client completes the Bruce treadmill protocol and has an estimated $\dot{V}O_{2max}$ of 12 METs. She weighs 47 kg.

32. If she cranks the arm ergometer against a 2.0 kg resistance at 50 rpm, at what percent $\dot{V}O_{2max}$ is she working?
    a) 50%
    b) 45%
    c) 55%
    d) 60%

33. If you want her to work at 50% of her $\dot{V}O_{2reserve}$ by stepping on a 15-inch step, what is her step rate?
    a) 15 steps/min
    b) 17 steps/min
    c) 20 steps/min
    d) 12 steps/min

34. Working at 7 METs, how many minutes must she exercise in order to expend 200 kcals?
    a) 35 minutes
    b) 60 minutes
    c) 20 minutes
    d) 42 minutes

ANSWERS

BRIEF ANSWERS

1. d	8. d	15. b	22. b	29. d
2. a	9. b	16. b	23. c	30. d
3. c	10. a	17. a	24. c	31. a
4. a	11. d	18. a	25. a	32. b
5. d	12. b	19. c	26. c	33. b
6. a	13. a	20. d	27. b	34. a
7. c	14. c	21. a	28. a	

DETAILED SOLUTIONS

1. Convert watts to kg · m/min.

$$20 \cdot 6.12 = 734.4$$

Use the leg ergometer equation to convert watts to $\dot{V}O_2$:

$$\dot{V}O_2 \text{ (mL/kg/min)} = 1.8 \cdot \frac{\text{kg} \cdot \text{m/min}}{\text{body mass (kg)}} + 7$$

$$= 1.8 \left(\frac{734}{75} \right) + 7$$

$$= 1.8 \cdot 9.8 + 7$$

$$= 17.6 + 7$$

$$= 24.6 \text{ mL/kg/min}$$

Convert to METs:

$$\frac{24.6 \text{ mL/kg/min}}{3.5} = \textbf{7 METs}$$

2. Convert $\dot{V}O_2$ (mL/kg/min) to L/min.

$$24.6 \text{ mL/kg/min} \cdot 75 = 1845 \text{ mL/min}$$

$$\frac{1845}{1000} = 1.845 \text{ L/min}$$

Convert to kcals/min:

$$1.845 \cdot 5 = 9.2 \text{ kcals/min}$$

Multiply by 30 minutes:

$$9.2 \cdot 30 = \textbf{276 kcals}$$

3. Use the $\dot{V}O_2$ from the leg ergometer work rate and use the walking equation to determine speed.

$$\dot{V}O_2 \ (mL/kg/min) = (0.1 \cdot S \ m/min) + (S \ m/min \cdot G \cdot 1.8) + 3.5$$
$$24.6 = (0.1 \cdot S \ m/min) + (S \ m/min \cdot 0.10 \cdot 1.8) + 3.5$$
$$-3.5 \qquad\qquad\qquad\qquad\qquad\qquad\qquad -3.5$$
$$21.1 = 0.1 \ S \ m/min + 0.18 \ S \ m/min$$
$$21.1 = 0.28 \ S \ m/min$$
$$\frac{21.1}{0.28} = \left(\frac{0.28}{0.28}\right) S \ m/min$$
$$75.4 = S \ m/min$$

Convert to mile/hr:

$$\frac{75.4}{26.8} = \mathbf{2.8 \ miles/hr}$$

4. Convert 9 inches to meters.

$$9 \cdot 2.54 = 22.86 \ cm$$
$$\frac{22.86}{100} = 0.23 \ m$$

Use the stepping equation to solve for $\dot{V}O_2$:

$$\dot{V}O_2 \ (mL/kg/min) = (0.2 \cdot f) + (1.33 \cdot 1.88 \cdot H \cdot f) + 3.5$$
$$= (0.2 \cdot 20) + (1.33 \cdot 1.88 \cdot 0.23 \cdot 20) + 3.5$$
$$= 4 + 11.5 + 3.5$$
$$= 18.5 \ mL/kg/min$$

Convert to L/min and then to kcals/min:

$$18.5 \cdot 58 \ kg = 1073 \ mL/min$$
$$\frac{1073}{1000} = 1.07 \ L/min$$
$$1.07 \cdot 5 = 5.35 \ kcals/min$$

Multiply by 20 minutes:

$$5.35 \cdot 20 = \mathbf{107 \ kcals}$$

5. Use $\dot{V}O_2$ in $mL \cdot kg^{-1} \cdot min^{-1}$ and the leg ergometer equation to determine work rate.

$$\dot{V}O_2 = 1.8 \left(\frac{kg \cdot m/min}{body \ mass \ (kg)}\right) + 7$$

$$18.5 = 1.8 \left(\frac{kg \cdot m/min}{58}\right) + 7$$

$$-7 \qquad\qquad\qquad\qquad\qquad -7$$

$$11.5 = 1.8 \left(\frac{\text{kg} \cdot \text{m/min}}{58} \right)$$

$$\cdot\, 58 = \cdot\, 58$$
$$667 = 1.8 \text{ kg} \cdot \text{m/min}$$
$$\frac{667}{1.8} = \frac{1.8}{1.8} \text{ kg} \cdot \text{m/min}$$
$$\mathbf{370 = kg \cdot m/min}$$

6. Convert mile/hr to m/min.

$$9.0 \cdot 26.8 = 241.2$$

Use running equation to determine $\dot{V}O_2$:

$$\dot{V}O_2 = (0.2 \cdot S \text{ m/min}) + (S \text{ m/min} \cdot G \cdot 0.9) + 3.5$$
$$= (0.2 \times 241.2) + (241.2 \cdot 0 \cdot 0.9) + 3.5$$
$$= 48.24 + 0 + 3.5$$
$$= \mathbf{51.74 \text{ mL/kg/min}}$$

7. Convert $\dot{V}O_2$ (mL/kg/min to L/min).

$$51.7 \cdot 80 = 4136 \text{ mL/min}$$

$$\frac{4136}{1000} = 4.1 \text{ L/min}$$

Convert to kcals/min:

$$4.1 \cdot 5 = \mathbf{20.5 \text{ kcals/min}}$$

8. Determine 40% of $\dot{V}O_{2max}$.

$$32.0 \cdot 0.40 = 12.8 \text{ mL/kg/min}$$

Use the arm ergometer equation to solve for work rate:

$$\dot{V}O_2 \text{ (mL/kg/min)} = 3.0 \left(\frac{\text{kg} \cdot \text{m/min}}{\text{body mass (kg)}} \right) + 3.5$$

$$12.8 = 3.0 \left(\frac{\text{kg} \cdot \text{m/min}}{74} \right) + 3.5$$

$$-3.5 \qquad\qquad\qquad -3.5$$

$$9.3 = 3.0 \left(\frac{\text{kg} \cdot \text{m/min}}{74} \right)$$

$$\cdot\, 74 = \cdot\, 74$$

$$688.2 = 3.0 \text{ kg} \cdot \text{m/min}$$

$$\frac{688.2}{3.0} = \frac{3.0}{3.0} \text{ kg} \cdot \text{m/min}$$

$$\mathbf{229.4 = kg \cdot m/min}$$

9. Determine 80% of $\dot{V}O_{2R}$.

$$\dot{V}O_{2R} = [(32.0 - 3.5) \cdot 0.80] + 3.5$$
$$= [28.5 \cdot 0.80] + 3.5$$
$$= 22.8 + 3.5$$
$$= 26.3 \text{ mL/kg/min}$$

Use leg ergometer equation to solve for work rate:

$$\dot{V}O_2 = 1.8 \left(\frac{\text{kg} \cdot \text{m/min}}{\text{body mass (kg)}} \right) + 7$$

$$26.3 = 1.8 \left(\frac{\text{kg} \cdot \text{m/min}}{74} \right) + 7$$

$$\frac{-7}{19.3} = 1.8 \left(\frac{\text{kg} \cdot \text{m/min}}{74} \right) \quad -7$$

$$\cdot 74 = \cdot 74$$

$$1428.2 = 1.8 \text{ kg} \cdot \text{m/min}$$

$$\frac{1428.2}{1.8} = \frac{1.8}{1.8} \text{ kg} \cdot \text{m/min}$$

$$\mathbf{793 = kg \cdot m/min}$$

10. Determine 60% of $\dot{V}O_{2max}$.

$$32.0 \cdot 0.60 = 19.2 \text{ mL/kg/min}$$

Convert to L/min:

$$19.2 \cdot 74 \text{ kg} = 1420.8 \text{ mL/min}$$

$$\frac{1420.8}{1000} = 1.4 \text{ L/min}$$

Convert to kcals/min:

$$1.42 \text{ L/min} \cdot 5 = \mathbf{7.1 \text{ kcals/min}}$$

11. Convert speed to m/min.

$$2.5 \text{ miles/hr} \cdot 26.8 = 67 \text{ m/min}$$

Convert grade to a fraction:

$$\frac{12\%}{100} = 0.12$$

Use the walking equation to solve for $\dot{V}O_2$:

$$\begin{aligned}
\dot{V}O_2 &= (0.1 \cdot S \text{ m/min}) + (S \text{ m/min} \cdot G \times 1.8) + 3.5 \\
&= (0.1 \cdot 67) + (67 \cdot 0.12 \cdot 1.8) + 3.5 \\
&= 6.7 + 14.47 + 3.5 \\
&= \mathbf{24.7 \ mL/kg/min}
\end{aligned}$$

12. Use the $\dot{V}O_2$ from the previous problem to solve for work rate using the leg ergometer equation.

$$\dot{V}O_2 = 1.8 \left(\frac{kg \cdot m/min}{body \ mass \ (kg)} \right) + 7$$

$$24.7 = 1.8 \left(\frac{kg \cdot m/min}{54} \right) + 7$$

$$\underline{-7} \qquad\qquad\qquad\qquad \underline{-7}$$

$$17.7 = 1.8 \left(\frac{kg \cdot m/min}{54} \right)$$

$$\cdot 54 = \cdot 54$$

$$955.8 = 1.8 \ kg \cdot m/min$$

$$\frac{955.8}{1.8} = \frac{1.8}{1.8} \ kg \cdot m/min$$

$$\mathbf{531 = kg \cdot m/min}$$

13. Convert step height to meters.

$$12 \text{ in} \cdot 0.0254 = 0.30 \text{ m}$$

Use the $\dot{V}O_2$ from the previous question and solve for step rate using the stepping equation.

$$\begin{aligned}
\dot{V}O_2 \ (mL/kg/min) &= (0.2 \cdot f) + (1.33 \cdot 1.88 \cdot H \cdot f) + 3.5 \\
24.7 \ mL/kg/min &= 0.2 \ f + (1.33 \cdot 1.88 \cdot 0.30 \cdot f) + 3.5
\end{aligned}$$

$$\underline{-3.5} \qquad\qquad\qquad\qquad\qquad\qquad \underline{-3.5}$$

$$21.2 = 0.2 \ f + 0.75 \ f$$

$$21.2 = 0.95 \ f$$

$$\frac{21.2}{0.95} = \frac{0.95}{0.95} \ f$$

$$\mathbf{22.3 \ steps/min = f}$$

14. Compute work rate in kg · m/min.

$$kg \cdot m/min = load \cdot m/rev \cdot rev/min$$
$$= 2.0 \cdot 6 \cdot 65$$
$$kg \cdot m/min = 780$$

Use the leg ergometer equation to solve for $\dot{V}O_2$:

$$\dot{V}O_2 \text{ (mL/kg/min)} = 1.8 \left(\frac{kg \cdot m/min}{body\ mass\ (kg)} \right) + 7$$

$$= 1.8 \left(\frac{780}{55} \right) + 7$$

$$= 1.8\,(14.2) + 7$$

$$= 32.6 \text{ mL/kg/min}$$

$$METs = \frac{32.6}{3.5} = 9.3$$

$$\dot{V}O_2 \text{ in L/min} =$$

$$32.6 \text{ mL/kg/min} \cdot 55 \text{ kg} = 1793 \text{ mL/min}$$

$$\frac{1793}{1000} = \textbf{1.8 L/min}$$

15. Convert $\dot{V}O_{2max}$ from L/min to mL/kg/min.

$$3.0 \cdot 1000 = 3000 \text{ mL/min}$$

$$\frac{3000}{55} = 54.5 \text{ mL/kg/min}$$

Determine 70% of $\dot{V}O_2$ R:

$$((54.5 - 3.5) \cdot 0.70) + 3.5$$
$$(51 \cdot 0.70) + 3.5$$
$$35.7 + 3.5$$
$$= 39.2 \text{ mL/kg/min}$$

Use the leg ergometer equation determine work rate:

$$\dot{V}O_2 = 1.8 \left(\frac{kg \cdot m/min}{body\ mass\ (kg)} \right) + 7$$

$$39.2 = 1.8 \left(\frac{kg \cdot m/min}{55} \right) + 7$$

$$-7 \qquad\qquad\qquad\qquad -7$$

$$32.2 = 1.8 \left(\frac{kg \cdot m/min}{55} \right)$$

$$\cdot 55 = \cdot 55$$
$$1771 = 1.8 \text{ kg} \cdot m/min$$

$$\frac{1771}{1.8} = \frac{1.8}{1.8} \text{ kg} \cdot \text{m/min}$$

$$984 = \text{kg} \cdot \text{m} \cdot \text{min}^{-1}$$

Convert to watts:

$$\frac{984}{6.12} = \textbf{160.78 watts}$$

16. Determine running speed in m/min.

$$6.5 \text{ miles/hr} \cdot 26.8 = 174.2 \text{ m/min}$$

Use the running equation to solve for $\dot{V}O_2$

$$
\begin{aligned}
\dot{V}O_2 \text{ (mL/kg/min)} &= (0.2 \cdot \text{S m/min}) + (\text{S m/min} \cdot \text{G} \cdot 0.9) + 3.5 \\
&= (0.2 \cdot 174.2) + (174.2 \cdot 0 \cdot 0.9) + 3.5 \\
&= 34.8 + 0 + 3.5 \\
&= 38.3 \text{ mL/kg/min}
\end{aligned}
$$

Convert to METs:

$$\frac{38.3}{3.5} = \textbf{11 METs}$$

17. Determine work rate (remember: arm ergometer flywheel is 2.4 m/rev).

$$
\begin{aligned}
\text{kg} \cdot \text{m/min} &= \text{Load} \cdot \text{m/rev} \cdot \text{rev/min} \\
&= 1.5 \cdot 2.4 \cdot 60 \\
&= 216 \text{ kg} \cdot \text{m/min}
\end{aligned}
$$

Use the arm ergometer equation to determine $\dot{V}O_2$:

$$
\begin{aligned}
\dot{V}O_2 \text{ (mL/kg/min)} &= 3.0 \left(\frac{\text{kg} \cdot \text{m/min}}{\text{body mass (kg)}} \right) + 3.5 \\
&= 3.0 \left(\frac{216}{85} \right) + 3.5 \\
&= 3.0 \, (2.5) + 3.5 \\
&= 7.5 + 3.5 \\
&= 11 \text{ mL/kg/min}
\end{aligned}
$$

Convert to L/min, then to kcals/min:

$$11.0 \text{ (mL/kg/min)} \cdot 85 = 935 \text{ mL/min}$$

$$\frac{935}{1000} = 0.935 \text{ L/min}$$

$$0.935 \cdot 5 = \textbf{4.7 kcals/min}$$

18. Convert step height to meters.

$$10 \text{ in} \cdot 0.0254 = 0.25 \text{ m}$$

Use the stepping equation to solve for $\dot{V}O_2$:

$$
\begin{aligned}
\dot{V}O_2 \text{ (mL/kg/min)} &= (0.2 \cdot f) + (1.33 \cdot 1.88 \cdot H \cdot f) + 3.5 \\
&= (0.2 \cdot 24) + (1.33 \cdot 1.88 \cdot 0.25 \cdot 24) + 3.5 \\
&= 4.8 + 15 + 3.5 \\
&= 23.3 \text{ mL/kg/min}
\end{aligned}
$$

Convert to L/min, then to kcals/min:

$$23.3 \text{ (mL/kg/min)} \cdot 85 \text{ kg} = 1980 \text{ mL/min}$$

$$\frac{1980}{1000} = 1.98 \text{ L/min}$$

$$1.98 \cdot 5 = \textbf{9.9 kcals/min}$$

19. Convert speed to m/min and grade to a fraction.

$$3.5 \text{ miles/hr} \cdot 26.8 = 93.8 \text{ m/min}$$

$$\frac{12\%}{100} = 0.12$$

Use the walking equation to solve for $\dot{V}O_2$:

$$
\begin{aligned}
\dot{V}O_2 \text{ (mL/kg/min)} &= (0.1 \cdot S \text{ m/min}) + (S \text{ m/min} \cdot G \cdot 1.8) + 3.5 \\
&= (0.1 \cdot 93.8) + (93.8 \cdot 0.12 \cdot 1.8) + 3.5 \\
&= 9.38 + 20.3 + 3.5 \\
&= 33.2 \text{ mL/kg/min}
\end{aligned}
$$

Convert to L/min, then to kcals/min:

$$33.2 \text{ mL/kg/min} \cdot 85 \text{ kg} = 2822 \text{ mL/min}$$

$$\frac{2822}{1000} = 2.82 \text{ L/min}$$

$$2.82 \text{ L/min} \cdot 5 = \textbf{14.1 kcals/min}$$

20. Determine 40% of MET capacity:

$$8 \text{ METs} \cdot 0.40 = 3.2 \text{ METs}$$

Convert to mL/kg/min:

$$3.2 \text{ METs} \cdot 3.5 = 11.2 \text{ mL/kg/min}$$

Use the arm ergometer equation to determine work rate:

$$\dot{V}O_2 = 3.0 \left(\frac{kg \cdot m/min}{body\ mass\ (kg)} \right) + 3.5$$

$$11.2 = 3.0 \left(\frac{kg \cdot m/min}{90} \right) + 3.5$$

$$-3.5 \qquad\qquad\qquad\qquad -3.5$$

$$7.7 = 3.0 \left(\frac{kg \cdot m/min}{90} \right)$$

$$\cdot\ 90 \qquad\qquad\qquad\qquad \cdot\ 90$$

$$693 = 3.0\ kg \cdot m/min$$

$$\frac{693}{3.0} = \frac{3.0}{3.0}\ kg \cdot m/min$$

$$\mathbf{231 = kg \cdot m/min}$$

21. Determine 65% of $\dot{V}O_2R$.

$$\dot{V}O_2R = ((8\ METs - 1\ METs) \cdot 0.65) + 1\ MET$$
$$= (7 \cdot 0.65) + 1$$
$$= 4.55 + 1$$
$$= 5.55\ METs$$

Convert to mL/kg/min:

$$5.55 \cdot 3.5 = 19.4\ mL/kg/min$$

Convert to L/min, then to kcals/min:

$$19.4\ mL/kg/min \cdot 90\ kg = 1746\ mL/min$$

$$\frac{1746}{1000} = 1.746\ L/min$$

$$1.746 \cdot 5 = \mathbf{8.7\ kcals/min}$$

22. Convert METs to mL/kg/min.

$$8\ METs \cdot 3.5 = 28.0\ mL/kg/min$$

Determine 50% of $\dot{V}O_{2max}$:

$$28.0 \cdot 0.50 = 14.0\ mL/kg/min$$

Convert speed to m/min:

$$1\ mile/hr \cdot 26.8 = 80.4\ m/min$$

Use the walking equation to solve for grade:

$$\dot{V}O_2 \text{ (mL/kg/min)} = (0.1 \cdot S \text{ m/min}) + (S \text{ m/min} \cdot G \cdot 1.8) + 3.5$$
$$14.0 \text{ mL/kg/min} = (0.1 \cdot 80.4) + (80.4 \cdot G \cdot 1.8) + 3.5$$

$$\begin{array}{lr} - 3.5 & - 3.5 \end{array}$$

$$10.5 = (8.04) + (144.7 \ G)$$

$$\begin{array}{lr} -8.04 & -8.04 \end{array}$$

$$2.46 = 144.7 \ G$$

$$\frac{2.46}{144.7} = \frac{144.7}{144.7} \ G$$

$$0.017 = G$$

Convert grade from decimal to percent:

$$0.017 \cdot 100 = 1.7 \ \%$$

23. Convert speed to m/min.

$$3.3 \text{ miles/hr} \cdot 26.8 = 88.4 \text{ m/min}$$

Use the walking equation to solve for $\dot{V}O_2$:

$$\dot{V}O_2 \text{ (mL/kg/min)} = (0.1 \cdot S \text{ m/min}) + (S \text{ m/min} \cdot G \cdot 1.8) + 3.5$$
$$= (0.1 \cdot 88.4) + (88.4 \cdot 0 \cdot 1.8) + 3.5$$
$$= 8.84 + 0 + 3.5$$
$$= 12.3 \text{ mL/kg/min}$$

Convert to L/min, then to kcals/min:

$$12.3 \text{ mL/kg/min} \cdot 75 \text{ kg} = 922.5 \text{ mL/min}$$

$$\frac{922.5}{1000} = 0.92 \text{ L/min}$$

$$0.92 \cdot 5 = 4.6 \text{ kcals/min}$$

Determine time to 300 kcals:

$$\frac{300 \text{ kcals}}{4.6 \text{ kcals/min}} = \textbf{65.2 min}$$

24. Convert kcals/min to L/min and then to mL/kg/min.

$$\frac{6.0}{5} = 1.2 \text{ L/min}$$

$$1.2 \cdot 1000 = 1200 \text{ mL/min}$$

$$\frac{1200}{75} = 16 \text{ mL/kg/min}$$

Use the leg ergometer equation to solve for work rate:

$$\dot{V}O_2 = 1.8 \left(\frac{\text{kg} \cdot \text{m/min}}{\text{body mass (kg)}} \right) + 7$$

$$16 = 1.8 \left(\frac{\text{kg} \cdot \text{m/min}}{75} \right) + 7$$

$$\begin{array}{cc} -7 & \quad\quad -7 \end{array}$$

$$9 = 1.8 \frac{\text{kg} \cdot \text{m/min}}{75}$$

$$\begin{array}{cc} \cdot\, 75 & \quad\quad \cdot\, 75 \end{array}$$

$$675 = 1.8 \cdot \text{kg} \cdot \text{m/min}$$

$$\frac{675}{1.8} = \frac{1.8}{1.8} \text{kg} \cdot \text{m/min}$$

$$375 = \text{kg} \cdot \text{m/min}$$

Convert to watts:

$$\frac{375}{6.12} = \textbf{61 watts}$$

25. Determine work rate in kg · m/min (remember, the arm ergometer flywheel is 2.4 m/rev).

$$\begin{aligned} \text{kg} \cdot \text{m/min} &= \text{Load} \cdot \text{m/rev} \cdot \text{rev/min} \\ &= 1.0 \cdot 2.4 \cdot 50 \\ &= 120 \text{ kg} \cdot \text{m/min} \end{aligned}$$

Use the arm ergometer equation to determine $\dot{V}O_2$:

$$\dot{V}O_2 \text{ (mL/kg/min)} = 3.0 \left(\frac{\text{kg} \cdot \text{m/min}}{\text{body mass (kg)}} \right) + 3.5$$

$$= 3.0 \left(\frac{120}{75} \right) + 3.5$$

$$= 3.0 \, (1.6) + 3.5$$

$$= 4.8 + 3.5$$

$$= 8.3 \text{ mL/kg/min}$$

Convert to L/min, then to kcals/min:

$$8.3 \text{ mL/kg/min} \cdot 75 = 622.5 \text{ mL/min}$$

$$\frac{622.5}{1000} = 0.623 \text{ L/min}$$

$$0.623 \cdot 5 = 3.1 \text{ kcals/min}$$

Determine how much time to attain 300 kcals:

300 kcals

3.1 kcals/min = **97 minutes (1 hour 37 minutes)**

26. Determine the $\dot{V}O_2$ at 80% of $\dot{V}O_{2max}$

$$73.0 \text{ mL/kg/min} \cdot 0.80 = 58.4 \text{ mL/kg/min}$$

Use the running equation to determine running speed (assume 0% grade):

$$\dot{V}O_2 = (0.2 \cdot S \text{ m/min}) + (S \text{ m/min} \cdot G \cdot 0.9) + 3.5$$
$$58.4 = 0.2 \text{ S m/min} + 0 + 3.5$$
$$-3.5 \qquad\qquad -3.5$$
$$57.9 = 0.2 \text{ S m/min}$$
$$\frac{57.9}{0.2} = \frac{0.2}{0.2} \text{ S m/min}$$
$$289.5 = S \text{ m/min}$$

Convert to miles/hr:

$$\frac{289.5}{26.8} = \textbf{10.8 miles/hr}$$

27. Use the leg ergometer equation to determine $\dot{V}O_2$.

$$\dot{V}O_2 \text{ (mL/kg/min)} = 1.8 \left(\frac{\text{kg} \cdot \text{m/min}}{\text{body mass (kg)}} \right) + 7$$
$$= 1.8 \left(\frac{1100}{73} \right) + 7$$
$$= 1.8 \,(15.07) + 7$$
$$= 27.1 + 7$$
$$= 34.1 \text{ mL/kg/min}$$

Determine relative intensity:

$$\left(\frac{\text{Exercise } \dot{V}O_2}{\dot{V}O_{2max}} \right) \times 100$$
$$\left(\frac{34.1}{73} \right) \cdot 100 = \textbf{46.7\%}$$

28. Use the running equation to determine speed. 5% grade = .05

$$\dot{V}O_2 = (0.2 \cdot S \text{ m/min}) + (S \text{ m/min} \cdot G \cdot 0.9) + 3.5$$
$$50 = 0.2 \cdot S \text{ m/min} + (S \text{ m/min} \cdot 0.05 \cdot 0.9) + 3.5$$
$$-3.5 \qquad\qquad\qquad\qquad\qquad -3.5$$

$$46.5 = 0.2 \cdot S \text{ m/min} + 0.045 \, S \text{ m/min}$$
$$46.5 = 0.245 \, S \text{ m/min}$$
$$\frac{46.5}{0.245} = \frac{0.245}{0.245}$$
190 = m/min

29. Determine work rate.

$$\text{kg} \cdot \text{m/min} = R \cdot \text{m/rev} \times \text{rev/min}$$
$$= 1.5 \cdot 6 \cdot 75$$
$$= 675 \text{ kg} \cdot \text{m/min}$$

Determine $\dot{V}O_2$ using the leg ergometer:

$$\dot{V}O_2 \text{ (mL/kg/min)} = 1.8 \left(\frac{\text{kg} \cdot \text{m/min}}{\text{body mass (kg)}} \right) + 7$$

$$= 1.8 \left(\frac{675}{55} \right) + 7$$

$$= 1.8 \, (12.3) + 7$$
$$= 22.1 + 7$$
$$= 29.1 \text{ mL/kg/min}$$

Convert to METs:

$$\frac{29.1}{3.5} = \textbf{8.3 METs}$$

30. Convert grade to a fraction.

$$\frac{10\%}{100} = 0.10$$

Convert METs to $\dot{V}O_2$:

$$8.3 \text{ METs} \cdot 3.5 = 29.1 \text{ mL/kg/min}$$

Use the walking equation to determine speed:

$$\dot{V}O_2 \text{(mL/kg/min)} = (0.1 \cdot S \text{ m/min}) + (S \text{ m/min} \cdot G \cdot 1.8) + 3.5$$
$$29.1 = (0.1 \, S \text{ m/min}) + (S \text{ m/min} \cdot 0.10 \cdot 1.8) + 3.5$$
$$-3.5 \qquad\qquad\qquad\qquad\qquad\qquad\qquad -3.5$$
$$25.6 = 0.1 \, S \text{ m/min} + 0.180 \, S \text{ m/min}$$
$$25.6 = 0.280 \, S \text{ m/min}$$

$$\frac{25.6}{0.280} = \frac{0.280}{0.280} \, S \text{ m/min}$$

$$91.4 = S \text{ m/min}$$

Convert speed to miles/hr:

$$\frac{91.4}{26.8} = \textbf{3.4 miles/hr}$$

31. Using the $\dot{V}O_2$ computed in question 20, determine $\dot{V}O_2$ in L/min, then kcals/min.

$$29.1 \text{ mL/kg/min} \cdot 55 \text{ kg} = 1600.5 \text{ mL/min}$$

$$\frac{1600.5}{1000} = 1.6 \text{ L/min}$$

$$1.6 \cdot 5 = 8.0 \text{ kcals/min}$$

Multiply by 30 minutes to compute the total:

$$8.0 \text{ kcals/min} \cdot 30 \text{ min} = \textbf{240 kcals}$$

32. Determine work rate (remember, the arm ergometer flywheel turns 2.4 m/rev).

$$\text{kg} \cdot \text{m/min} = \text{R} \cdot \text{m/rev} \cdot \text{rev/min}$$
$$= 2.0 \cdot 2.4 \cdot 50$$
$$= 240 \text{ kg} \cdot \text{m/min}$$

Determine $\dot{V}O_2$ using the arm ergometer equation:

$$\dot{V}O_2 = 3.0 \left(\frac{\text{kg} \cdot \text{m/min}}{\text{body mass (kg)}} \right) + 3.5$$

$$= 3.0 \left(\frac{240}{47} \right) + 3.5$$

$$= 3.0 \,(5.1) + 3.5$$
$$= 15.3 + 3.5$$
$$= 18.8 \text{ mL/kg/min}$$

Convert to METs:

$$\frac{18.8}{3.5} = 5.4 \text{ METs}$$

Determine relative intensity:

$$\left(\frac{5.4 \text{ METs}}{12 \text{ METs}} \right) \cdot 100 = \textbf{45\% of } \dot{V}O_{2max}$$

33. Determine 50% of $\dot{V}O_{2Reserve}$.

$$\dot{V}O_{2R} = ((12 \text{ METs} - 1 \text{ MET}) \cdot 0.50) + 1 \text{ MET}$$
$$= (11 \cdot 0.50) + 1$$

$$= 5.5 + 1$$
$$= 6.5 \text{ METs}$$
$$6.54 \cdot 3.5 = 22.9 \text{ mL/kg/min}$$

Convert step equation and the $\dot{V}O_2$ above to solve for step rate:

$$\dot{V}O_2 \text{ mL/kg/min} = (0.2 \cdot f) + (1.33 \cdot 1.88 \cdot H \cdot f) + 3.5$$
$$22.9 = 0.2 f + (1.33 \cdot 1.88 \cdot 0.38 \cdot f) + 3.5$$
$$\underline{-3.5} \qquad\qquad\qquad\qquad\qquad \underline{-3.5}$$
$$19.4 = 0.2 f + 0.95 f$$
$$19.4 = 1.15 f$$
$$\frac{19.4}{1.15} = \frac{1.15}{1.15} f$$

16.9 steps/min = f

34. Convert METs to $\dot{V}O_2$ mL/kg/min then L/min, then kcals/min.

$$7 \text{ METs} \cdot 3.5 = 24.5 \text{ mL/kg/min}$$
$$24.5 \text{ mL/kg/min} \cdot 47 \text{ kg} = 1151.5 \text{ mL/min}$$
$$\frac{1151.5}{1000} = 1.15 \text{ L/min}$$
$$1.15 \cdot 5 = 5.8 \text{ kcals/min}$$

Determine time to expend 200 kcals:

$$\frac{200 \text{ kcals}}{5.8 \text{ kcals/min}} = \textbf{34.5 minutes}$$

SAMPLE MULTIPLE CHOICE

METs = Miles/Hr % Grade

The following questions are similar in type and difficulty to the questions you may encounter on the certification examinations of the American College of Sports Medicine. Included in this chapter are multiple coice questions that are singular or stand-alone as well as in a case study format. Also included are the answers to the questions at the end of this chapter.

SINGULAR OR STAND-ALONE TEST QUESTIONS

1. At an average exercise intensity of 10 METs, the energ expended in 30 minutes by a 110 lb (50 kg) woman is approximately _____ kcal.
   a) 232
   b) 262
   c) 300
   d) 332

2. A 35-year-old, 135 lb (61.4 kg) woman is pedaling on a Monark cycle ergometer at 100 watts (600 kg · m/min). During 30 minutes of exercise she willexpend a total of approximately _____ U. S. kcal.
   a) 181
   b) 226
   c) 297
   d) 329

3. What is the workload of a 165 lb (75 kg) person pedaling on a Monark cycle ergometer at 50 revolutions per minute at a resistance of 1.5 kg?
   a) 175 kg · m/min
   b) 450 kg · m/min
   c) 650 kg · m/min
   d) none of the above

4. A client weighs 201 lb (91.5 kg) and has a maximum aerobic power of 40.7 mL/kg/min. What exercise intensity (power output) on a cycle ergometer will enable him to maintain a target training intensity of 7.9 METs?
    a) 950 kg · m/min
    b) 1050 kg · m/min
    c) 1150 kg · m/min
    d) 1250 kg · m/min

5. Two persons, one weighing 50 kg and the other weighing 80 kg, have maximal oxygen consumptions of 52 mL/kg/min. They both exercise on a cycle ergometer at a power output of 100 watts. Which variable will show the least difference between the two subjects?
    a) METs
    b) kcal/min
    c) percent of maximal oxygen uptake
    d) $\dot{V}O_2$ (mL/kg/min)

6. If two persons, one weighing 50 kg and the other 70 kg, walk at 2.5 miles/hr and 12% grade on the tradmill, then which of the following variables will be approximately the same for each of the two subjects?
    a) METs
    b) kcal/min
    c) oxygen pulse $\left(\dfrac{\dot{V}O_2}{HR}\right)$
    d) $\dot{V}O_2$ (L/min)

7. To predict the oxygen requirements in METs for pedaling a cycle ergometer at a fixed itensity one must know the subject's:
    a) age and sex
    b) body weight
    c) heart rate
    d) both A and B

8. Sam weighs 67 kg and is running on the treadmill at 5.5 miles/hr up a 5% grade. What is Sam's energy expenditure in terms of kcals/min?
    a) 8.7 kcals/mn
    b) 10.2 kcals/min
    c) 13.3 kcals/min
    d) to calculate you need to know how many minutes he is exercising for

9. The concept of $\dot{V}O_{2reserve}$ ($\dot{V}O_{2R}$) has been employed in:
    a) exercise prescription
    b) energy expenditure

c) calculation of oxygen pulse

d) calclation of basal metabolic rate from heart rate

10. A subject wishes to lose 1 pound a week; half with exercise, half by reduced caloric intake. He weighs 185 pounds. He can exercise comfortably at 10.5 METs and has time to exercise 4 times per week. Ho long should he exercise in each session to lose the 1/2-pound a week from exercise?

a) 28.3 minutes per session

b) 30.0 minutes per session

c) 37.2 minutes per session

d) 41.2 minutes per session

11. What is the oxygen uptake of a person who weighs 64 kg and pedals on a Monark cycle ergometer at a resistance of 3.0 kg and a rate of 75 rpm?

a) 2878 mL/min

b) 2924 mL/min

c) 3150 mL/min

d) 3320 mL/min

12. If a subject exercises on a Monark at 2 kg and 80 rev/min and weighs 138 pounds, approximately how many kcal will be expended in 20 minutes?

a) 118 kcal

b) 216 kcal

c) 329 kcal

d) 415 kcal

13. If a subject has a functional capacity of 12 METs and can walk at 3.6 miles/hr, where would you set the grade so that she can work at 75% of her functional capacity?

a) 5.8%

b) 8.7%

c) 10.6%

d) 12.8%

14. If a subject has a $\dot{V}O_{2max}$ of 10 METs, and the exercise prescription calls for a training intensity of 6 METs, what should the pace per mile be (assume 5% grade)?

a) 12.2 minutes per mile

b) 14.7 inutes per mile

c) 15.3 minutes per mile

d) 17.4 minutes per mile

15. A 50 kg woman is exercising at a work intensity of 8 METs for 20 minutes. Approximately how many kcal will she expend during this time?
    a) 100 kcals
    b) 140 kcals
    c) 180 kcals
    d) 220 kcals

16. The oxygen cost of running on the level at 200 m/min would be about:
    a) 30 mL/kg/min
    b) 50 mL/kg/min
    c) 10 METs
    d) 12 METs

17. When cycling on a Monark leg ergometer, if a person has a $\dot{V}O_{2max}$ of 3.5 L/min, and the exercise prescription calls for an intensity of 60% $\dot{V}O_{2max}$, what should the workload be set at if the subject weighs 70 kilograms?
    a) 600 kg · m/min
    b) 750 kg · m/min
    c) 900 kg · m/min
    d) 1000 kg · m/min

18. In the above problem (#17), if the pedal rev/min are set at 50, what would the resistance be set at?
    a) 2.0 kg
    b) 2.5 kg
    c) 3.0 kg
    d) 3.5 kg

19. If an exercise prescription calls for a subject to run at 85% of $\dot{V}O_{2max}$ on a treadmill, what would you set the grade at if the speed is fixed at 5.0 mph and his $\dot{V}O_{2max}$ is 51 mL/g/min:
    a) 7.9%
    b) 9.4%
    c) 10.2%
    d) 11%

20. What is the oxygen cost, in METs, for Sally (125 pounds), who is stepping at a frequency of 20 steps/min on a step bench that is 8 inches high?
    a) 2.4
    b) 4.9
    c) 6.6
    d) 8.2

CASE STUDY QUESTIONS

The following two case studies each have several multiple choice questions associated with metabolic calculations.

CASE STUDY I

The following exercise data were collected on Mr. Jones, who is age 50 and weighs 75 kg, during a submaximal GXT on a treadmill.

Time (min)	Speed (miles/hr)	Grade (%)	HR (beats/min)	BP (mm Hg)	RPE (6–20 scale)
REST			72	120/80	
0–2	3.0	0.0	98	160/90	9
2–4	3.0	2.5	112	168/90	11
4–6	3.0	5.0	127	174/88	12
6–8	3.0	7.5	135	185/86	13
8–10	3.0	10.0	148	194/84	15
10–12	3.0	12.5	156	200/84	16

His $\dot{V}O_{2max}$ was predicted to be 44.5 mL/kg/min.

21. If you want Mr. Jones to exercise at 65% of his maximal aerobic capacity, what workload (speed, in mph) would you use if he were to walk on the treadmill up a 10% grade?
   a) 2.9 miles/hr
   b) 3.1 miles/hr
   c) 3.4 miles/hr
   d) not enough information to calculate

22. If Mr. Jones were to exercise at that workload (in question #21) for 45 minutes, how many calories would he expend?
   a) 300 kcals
   b) 378 kcals
   c) 415 kcals
   d) 455 kcals

23. Mr. Jones would like to lose 10 pounds and wants to know how long it might take him (in weeks) to expend the calories equivalent to a 10-pound loss if he exercised at that workload (in question #21) 5 times per week for 45 minutes?
   a) 12.7 weeks
   b) 14.5 weeks

c) 16.2 weeks

d) There is no way to solve for this answer with the information given.

24. A target heart rate using 75% of intensity, using the Karvonen formula, would be:

a) 128 beats/min

b) 139 beats/min

c) 145 beats/min

d) 152 beats/min

25. Looking at the submaximal graded exercise test results, what might be his oxygen cost (in mL/kg/min) for the workload that most closely matches this heart rate (from question 24)?

a) 3.0 miles/hr; 2.5% grade

b) 3.0 miles/hr; 5.0% grade

c) 3.0 miles/hr; 7.5 grade

d) 3.0 miles/hr; 10% grade

CASE STUDY II

Mrs. Barnes' GXT on a leg ergometer was conducted in your testing facilities without a physician present and with no ECG monitoring. Mrs. Barnes is 48 years old, weighs 148 pounds and has no major risk factors for coronary artery disease. The data collected during her test are presented below.

Time (min)	Work rate (kg · m/min)	Heart rate (beats/min)	Blood pressure (mm Hg)	RPE (6–20 scale)
REST		68	116/78	
0–2	300	104	144/92	10
2–4	450	118	152/100	11
4–6	600	129	164/106	12
6–8	750	136	172/110	14
8–10	900	142	188/112	16

26. Mrs. Barnes thought that with the cycle exercise workload of 750 kg · m/min, her heart rate response to that workload would be 136 beats/min. At what percentage (using the Karvonen formula) of intensity would that be?

a) 55%

b) 60%

c) 65%

d) 70%

27. If Mrs. Barnes exercises at this workload (750 kg · m/min), what would be her energy expenditure in kcals/min?
    a) 9.1 kcals/min
    b) 9.9 kcals/min
    c) 10.5 kcals/min
    d) not enough information to calculate

28. Given this workload (750 kg · m/min), how long, in minutes, would she have to ride the cycle to expend 200 kcals in a session?
    a) 17.2 minutes
    b) 21.9 minutes
    c) 27.5 minutes
    d) 30.8 minutes

ANSWERS

1.	b
2.	b
3.	b
4.	b
5.	b
6.	a
7.	b
8.	c
9.	a
10.	b
11.	a
12.	b
13.	c
14.	d
15.	b
16.	d
17.	c
18.	c
19.	d
20.	b
21.	c
22.	d
23.	b
24.	c
25.	d
26.	c
27.	d
28.	b

METABOLIC EQUATIONS SUMMARY SHEET

METABOLIC EQUATIONS (S = SPEED, G = GRADE IN DECIMAL FORM)

Walking $\dot{V}O_2$ (mL/kg/min) = [S (m/min) · 0.1] + [S (m/min) · G · 1.8] + 3.5 mL/kg/min

Running $\dot{V}O_2$ (mL/kg/min) = [S m/min · 0.2] + [S m/min · G × 0.9] + 3.5

Leg Ergometer $\dot{V}O_2$ (mL/kg/min) = $1.8 \dfrac{\text{work rate (kg · m/min)}}{\text{body mass (kg)}} + 7$

Arm Ergometer $\dot{V}O_2$ (mL/kg/min) = $3.0 \dfrac{\text{kg · m/min}}{\text{body mass (kg)}} + 3.5$

Stepping $\dot{V}O_2$ (mL/kg/min) = (0.2 · f) + (1.33 · 1.88 · H · f) + 3.5

f = stepping rate   H = step height in meters

ESSENTIAL CONVERSIONS

Action	Conversion
Conversion of miles per hour to meters per minute	Miles/hr $\times$ 26.8
Conversion of percent grade to grade as fraction	$\dfrac{\% \text{ grade}}{100}$
Conversion of $\dot{V}O_2$ in mL/kg/min to METs	$\dfrac{\dot{V}O_2 \,(mL/kg/min)}{3.5}$
Conversion of $\dot{V}O_2$ in mL/kg/min to mL/min	$\dot{V}O_2 \,(mL/kg/min) \times$ Body Mass in KG
Conversion of $\dot{V}O_2$ in mL/min to L/min	$\dfrac{\dot{V}O_2 \, mL/min}{1000}$
Conversion of $\dot{V}O_2$ in L/min to kcals/min	$\dot{V}O_2 \, L/min \cdot 5$
Watts to kg · m/min conversion	$watt = \dfrac{kg \cdot m/min}{6.12}$
$\dfrac{kg \cdot m}{min}$	Resistance (kg) · m/rev · rev/min
Inches to centimeters conversion	Inches · 2.54
Centimeters to meters conversion	$\dfrac{Centimeters}{100}$
Pounds to kilograms conversion	$\dfrac{Pounds}{2.2}$

SELECTED TABLES FROM *ACSM's* GUIDELINES FOR EXERCISE TESTING AND PREPARATION, SEVENTH EDITION

TABLE D-2 FROM 7TH EDITION ACSM GUIDELINES
Approximate Energy Requirements in METs for Horizontal and Grade Walking

	mph	1.7	2.0	2.5	3.0	3.4	3.75
% Grade	m · min^{-1}	45.6	53.6	67.0	80.4	91.2	100.5
0		2.3	2.5	2.9	3.3	3.6	3.9
2.5		2.9	3.2	3.8	4.3	4.8	5.2
5.0		3.5	3.9	4.6	5.4	5.9	6.5
7.5		4.1	4.6	5.5	6.4	7.1	7.8
10.0		4.6	5.3	6.3	7.4	8.3	9.1
12.5		5.2	6.0	7.2	8.5	9.5	10.4
15.0		5.8	6.6	8.1	9.5	10.6	11.7
17.5		6.4	7.3	8.9	10.5	11.8	12.9
20.0		7.0	8.0	9.8	11.6	13.0	14.2
22.5		7.6	8.7	10.6	12.6	14.2	15.5
25.0		8.2	9.4	11.5	13.6	15.3	16.8

Reprinted from *ACSM's Guidelines for Exercise Testing and Prescription*. 7th ed. Indianapolis: American College of Sports Medicine, 2006.

TABLE D-3 FROM 7TH EDITION ACSM GUIDELINES
Approximate Energy Requirements in METs for Horizontal and Grade Jogging/Running

	mph	5	6	7	7.5	8	9	10
% Grade	m · min^{-1}	134	161	188	201	214	241	268
0		8.6	10.2	11.7	12.5	13.3	14.8	16.3
2.5		9.5	11.2	12.9	13.8	14.7	16.3	18.0
5.0		10.3	12.3	14.1	15.1	16.1	17.9	19.7
7.5		11.2	13.3	15.3	16.4	17.4	19.4	
10.0		12.0	14.3	16.5	17.7	18.8		
12.5		12.9	15.4	17.7	19.0			
15.0		13.8	16.4	18.9				

Reprinted from *ACSM's Guidelines for Exercise Testing and Prescription*. 7th ed. Indianapolis: American College of Sports Medicine, 2006.

TABLE D-4 FROM 7TH EDITION ACSM GUIDELINES
Approximate Energy Requirements in METs for Leg Ergometry

Body Wt.		Power Output (kg · m · mm^{-1} and W)						
kg	lb	300 50	450 75	600 100	750 125	900 150	1,050 175	1,200 (kg· m · min^{-1}) 200 (W)
50	110	5.1	6.6	8.2	9.7	11.3	12.8	14.3
60	132	4.6	5.9	7.1	8.4	9.7	11.0	12.3
70	154	4.2	5.3	6.4	7.5	8.6	9.7	10.8
80	176	3.9	4.9	5.9	6.8	7.8	8.8	9.7
90	198	3.7	4.6	5.4	6.3	7.1	8.0	8.9
100	220	3.5	4.3	5.1	5.9	6.6	7.4	8.2

Reprinted from *ACSM's Guidelines for Exercise Testing and Prescription*. 7th ed. Indianapolis: American College of Sports Medicine, 2006.

TABLE D-5 FROM 7TH EDITION ACSM GUIDELINES
Approximate Energy Requirements in METs for Arm Ergometry

Body Wt.		Power Output (kg · m · mm^{-1} and W)					
kg	lb	150 25	300 50	450 75	600 100	750 125	900 (kg · m · min^{-1}) 150 (W)
50	110	3.6	6.1	8.7	11.3	13.9	16.4
60	132	3.1	5.3	7.4	9.6	11.7	13.9
70	154	2.8	4.7	6.5	8.3	10.2	12.0
80	176	2.6	4.2	5.8	7.4	9.0	10.6
90	198	2.4	3.9	5.3	6.7	8.1	9.6
100	220	2.3	3.6	4.9	6.1	7.4	8.7

Reprinted from *ACSM's Guidelines for Exercise Testing and Prescription*. 7th ed. Indianapolis: American College of Sports Medicine, 2006.

TABLE D-6 FROM 7TH EDITION ACSM GUIDELINES
Approximate Energy Requirements in METs for Stair Stepping

Step Height		Stepping Rate per Minute					
in	m	20	22	24	26	28	30
4	0.102	3.5	3.8	4.0	4.3	4.5	4.8
6	0.152	4.2	4.6	4.9	5.2	5.5	5.8
8	0.203	4.9	5.3	5.7	6.1	6.5	6.9
10	0.254	5.6	6.1	6.5	7.0	7.5	7.9
12	0.305	6.3	6.8	7.4	7.9	8.4	9.0
14	0.356	7.0	7.6	8.2	8.8	9.4	10.0
16	0.406	7.7	8.4	9.0	9.7	10.4	11.1
18	0.457	8.4	9.1	9.9	10.6	11.4	12.1

Reprinted from *ACSM's Guidelines for Exercise Testing and Prescription*. 7th ed. Indianapolis: American College of Sports Medicine, 2006.

Andersen RE, Wadden TA. Validation of a cycle ergometry equation for predicting steady-rate $\dot{V}O_2$ in obese women. Med Sci Sports Exerc. 1995 Oct;27(10):1457–1460.

Balogun JA, Martin DA, Clendenin MA. Human energy expenditure during level walking on a treadmill at speeds of 54–130 m min^{-1}. Int Disabil Stud 1989;11:71–74.

Bastien GJ, Willems PA, Schepens B, et al. Effect of load and speed on the energetic cost of human walking. Eur J Appl Physiol 2005;94:76–83. Epub 2005 Jan 14.

Brisswalter J, Mottet D. Energy cost and stride duration variability at preferred transition gait speed between walking and running. Can J Appl Physiol. 1996;21(6):471–480.

Bunc V, Dlouha R. Energy cost of treadmill walking. J Sports Med Phys Fitness. 1997;37:103–109.

Fellingham GW, Roundy ES, Fisher AG, Bryce GR. Caloric cost of walking and running. Med Sci Sports Exerc 1978;10(2):132–136.

Greiwe JS, Kohrt WM. Energy expenditure during walking and jogging. J Sports Med Phys Fitness 2000;40:297–302.

Hall C, Figueroa A, Fernhall B, et al. Energy expenditure of walking and running: comparison with prediction equations. Med Sci Sports Exerc 2004;36:2128–2134.

Jones AM, Doust JH. A 1% treadmill grade most accurately reflects the energetic cost of outdoor running. J Sports Sci 1996 Aug;14(4):321–327.

Lang PB, Latin RW, Berg KE, et al. The accuracy of the ACSM cycle ergometry equation. Med Sci Sports Exerc 1992;24:272–276.

Latin RW, Berg K, Kissinger K, et al. The accuracy of the ACSM stair-stepping equation. Med Sci Sports Exerc 2001 Oct;33(10):1785–1788.

Latin RW, Berg KE. The accuracy of the ACSM and a new cycle ergometry equation for young women. Med Sci Sports Exerc 1994;26:642–646.

Latin RW, Berg KE, Smith PR, Tolle P, Woodby-Brown S. Validation of a cycle ergometry equation for predicting steady-rate $\dot{V}O_2$. Med Sci Sports Exerc 1993;25:970–974.

Leger L, Mercier D. Gross energy cost of horizontal treadmill and track running. Sports Med. 1984;1:270–277.

Milani J, Fernhall B, Manfredi T. Estimating oxygen consumption during treadmill and arm ergometry activity in males with coronary artery disease. J Cardiopulm Rehabil 1996;16:394–401.

Minetti AE, Moia C, Roi GS, et al. Energy cost of walking and running at extreme uphill and downhill slopes. J Appl Physiol 2002;93:1039–1046.

Roberts RA, Wagner DR, Skemp KM. Oxygen consumption and energy expenditure of level versus downhill running. J Sports Med Phys Fitness. 1997;37:168–174.

Stanforth PR, Ruthven MD, Gagnon J, et al. Accuracy of prediction equations to estimate submaximal $\dot{V}O_2$ during cycle ergometry: the HERITAGE Family Study. Med Sci Sports Exerc 1999;31:183–188.

Stanforth D, Stanforth PR, Velasquez KS. Aerobic requirement of bench stepping. Int J Sports Med 1993;14:129–133.

Swain DP, Parrott JA, Bennett AR, et al. Validation of a new method for estimating $\dot{V}O_{2max}$ based on $\dot{V}O_2$ reserve. Med Sci Sports Exerc 2004;36:1421–1426.

Swan PD, Byrnes WC, Haymes EM. Energy expenditure estimates of the Caltrac accelerometer for running, race walking, and stepping. Br J Sports Med 1997;31:235–239.

Workman JM, Armstrong BW. Metabolic cost of walking: equation and model. J Appl Physiol 1986;61:1369–1374.

INDEX

Page numbers followed by *f* refer to illustrations; page numbers followed by *t* refer to tables.